TRANSFORMING RECONFIGURABLE SYSTEMS

A Festschrift Celebrating the 60th Birthday of Professor Peter Cheung

TRANSFORMING RECONFIGURABLE SYSTEMS

A Festschrift Celebrating the 60th Birthday of Professor Peter Cheung

Editors

Wayne Luk • George A. Constantinides

Imperial College London, UK

Imperial College Press

ICP

Published by

Imperial College Press
57 Shelton Street
Covent Garden
London WC2H 9HE

Distributed by

World Scientific Publishing Co. Pte. Ltd.
5 Toh Tuck Link, Singapore 596224
USA office: 27 Warren Street, Suite 401-402, Hackensack, NJ 07601
UK office: 57 Shelton Street, Covent Garden, London WC2H 9HE

Library of Congress Cataloging-in-Publication Data
Transforming reconfigurable systems : a festschrift celebrating the 60th birthday of professor Peter
Cheung / [compiled by] Wayne Luk, Imperial College London, UK, George A Constantinides,
Imperial College London, UK.
    pages cm
    "Technical papers based on presentations at a workshop at Imperial College London in May 2013
celebrating Professor Cheung's 60th birthday"--Preface.
    Includes bibliographical references and index.
    ISBN 978-1-78326-696-8 (hardcover : alk. paper)
    1. Adaptive computing systems--Congresses. 2. Electronic systems--Congresses. I. Luk, Wayne.
II. Constantinides, George A., 1975– III. Cheung, Peter Yee-Tung, 1954– IV. Imperial College,
London.
    QA76.9.A3T73 2015
    004--dc23

                    2014047056

British Library Cataloguing-in-Publication Data
A catalogue record for this book is available from the British Library.

Typeset by Stallion Press
Email: enquiries@stallionpress.com

Printed in Singapore

Preface

Over the last three decades, Professor Peter Cheung has made significant contributions to a variety of areas, such as analogue and digital computer-aided design tools, high-level synthesis and hardware/software codesign, low-power and high-performance circuit architectures for signal and image processing, and mixed-signal integrated-circuit design.

The area that has attracted his greatest attention, however, is reconfigurable systems and their design. His work has contributed to the transformation of this important and exciting discipline. For example, he developed the first dedicated multiplier unit for reconfigurable architectures; he pioneered a reconfigurable computer customized for professional video applications; and he is still making seminal contributions in addressing reliability challenges to field-programmable technology and reconfigurable systems.

His intellectual progeny include the better part of fifty research students and research associates, including one of the editors of this volume. Many of them have now become leaders in their field. Peter's most enduring impact will include not only those who have been fortunate enough to be directly inspired or mentored by him, but also his grand-students and great grand-students who indirectly benefit from his heritage.

Except for three years at Hewlett Packard, Peter has devoted his professional career to Imperial College, and has served with distinction as the Head of Department of Electrical and Electronic Engineering for several years. His outstanding capability and his loyalty to Imperial College in general, and to the Department of

Electrical and Electronic Engineering in particular, are legendary. For his department, Peter has made tremendous strides in ensuring excellence in both research and teaching, and in establishing sound governance and strong financial endowment; but above all, he has made his department a wonderful place to work and to study. His efforts have been rewarded by the warmth with which he is regarded by his colleagues and students.

Most of the papers in this festschrift are based on the presentations at a workshop on 3 May 2013 celebrating Peter's 60th birthday. We thank the contributors to this volume; while the topics covered vary significantly, all were able to relate their work to the contributions made by Peter. There are many more, especially most of Peter's former students, who wished to contribute than the limited space can offer, and we apologise for the lack of space. The effort of Thomas Chau, Eddie Hung and Tim Todman in assisting the production of this volume is much appreciated.

Last but by no means least: Happy Birthday, Peter! We look forward to working with you for many years to come!

Wayne Luk and George Constantinides

List of Contributors

Norbert Abel	Goethe-University Frankfurt
Samuel Bayliss	Imperial College London
David Boland	Imperial College London
Srinivas Boppu	University of Erlangen-Nuremberg
Andrew Brown	University of Southampton
Jason Cong	University of California, Los Angeles
George A. Constantinides	Imperial College London
Heiko Engel	Goethe-University Frankfurt
Michael J. Flynn	Maxeler Technologies and Stanford University
Paul J. Fox	University of Cambridge
Michael Frechtling	The University of Sydney
Steve Furber	University of Manchester
Jano Gebelein	Goethe-University Frankfurt
Frank Hannig	University of Erlangen-Nuremberg
Udo Kebschull	Goethe-University Frankfurt
Vahid Lari	University of Erlangen-Nuremberg
Philip H.W. Leong	The University of Sydney
Wayne Luk	Imperial College London
Sebastian Manz	Goethe-University Frankfurt
A. Theodore Markettos	University of Cambridge
Oskar Mencer	Maxeler Technologies and Imperial College London

Simon W. Moore	University of Cambridge
Michael Munday	Maxeler Technologies
Matthew Naylor	University of Cambridge
Oliver Pell	Maxeler Technologies
Lesley Shannon	Simon Fraser University
Jürgen Teich	University of Erlangen-Nuremberg
David B. Thomas	Imperial College London
Steve Wilton	University of British Columbia
Alex Yakovlev	Newcastle University

Table of Contents

Chapter 1

Accelerator-Rich Architectures — Computing Beyond Processors

Jason Cong

Computer Science Department,
University of California, Los Angeles

In order to drastically improve energy efficiency, we believe that future computer processors need to go beyond parallelization and provide architecture support of customization and specialization, enabling processor architectures to be adapted and optimized for different application domains. In particular, we believe that future processor architectures will make extensive use of accelerators to further increase energy efficiency. Such architectures present many new challenges and opportunities, such as accelerator synthesis, scheduling, sharing, virtualization, memory hierarchy optimization, and efficient compilation and runtime support. In this paper, I shall highlight some of our ongoing research in these areas that has taken place in the Center for Domain-Specific Computing (supported by the NSF Expeditions in Computing award). The material here is based on a talk that I presented in July 2012 at Imperial College London, hosted by Professor Peter Cheung.

1.1. Introduction

In order to meet today's ever-increasing computing needs and overcome power density limitations, the computing industry has halted simple processor frequency scaling and entered the era of parallelization, with tens to hundreds of computing cores integrated in a single processor, and hundreds to thousands of computing servers connected in a warehouse-scale data center. However, such highly parallel, general-purpose computing systems still face serious challenges in terms of performance, power, heat dissipation, space, and cost. We believe that we need to look beyond parallelization and focus on domain-specific customization to provide capabilities that adapt architecture to application in order to achieve significant power-performance efficiency improvement.

In fact, the performance gap between a totally customized solution (using an application-specific integrated circuit (ASIC)) and a general-purpose solution can be very large. A case study of the 128-bit key AES encryption algorithm was presented in [Ref. 1]. An ASIC implementation in $0.18\,\mu$m CMOS achieves $3.86\,$Gbit/s at $350\,$mW, while the same algorithm coded in Java and executed on an embedded SPARC processor yields $450\,$bit/s at $120\,$mW. This difference implies a performance/energy efficiency (measured in Gbit/s/W) gap of roughly 3 million! Therefore, one way to significantly improve the performance/energy efficiency is to have as much computation done as possible in accelerators designed in ASIC, instead of using general-purpose cores.

One argument against using accelerators is their low utilization and narrow workload coverage; however low utilization is no longer a serious problem. Given the utilization wall [Ref. 2] and dark silicon problem [Ref. 3] revealed in recent studies, we can activate only a fraction of computing elements on-chip at one time in future technologies, given the tight power and thermal budget. The problem of narrow workload coverage is addressed by using the composable accelerators or reconfigurable accelerators discussed below. Therefore, we believe that the future of processor architecture should be rich in accelerators, as opposed to having many cores.

In this paper, I shall highlight the progress made in the Center for Domain-Specific Computing (CDSC) [Ref. 4] on developing energy-efficient accelerator-rich architectures. I had the pleasure of a month-long visit to Imperial College London in July 2012, hosted by Professor Wayne Luk and supported by the Distinguished Visiting Fellow Program of the Royal Academy of Engineering. Professor Peter Cheung was very kind to let me use his office during my visit. The material covered here is largely based on an invited talk that I gave during my visit to the EEE Department.

The talk was hosted by Professor Cheung, who has made many fundamental contributions to the technology on which accelerator architectures can be based. I highlighted our progress in four areas: accelerator sharing and management (Sec. 1.2), memory support for accelerator-rich architectures (Sec. 1.3), on-chip communication for accelerator-rich architectures (Sec. 1.4), and software support for accelerator-rich architectures (Sec. 1.6). I also included some of the latest results on using fine-grain programmable fabrics to support composable accelerators (Sec. 1.2) and a recent effort on prototyping of accelerator-rich architectures (Sec. 1.5).

1.2. Accelerator Sharing and Management in Accelerator-Rich CMPs

We began our investigation of accelerator-rich architectures in 2010 and developed three generations of architecture templates. The first generation of architecture focuses on hardware support for accelerator-rich CMPs (ARC) [Ref. 5]. Figure 1.1 shows the overall architecture of ARC, which is composed of cores, accelerators, the

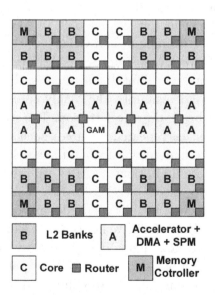

Fig. 1.1. Overall architecture of ARC (from [Ref. 5]).

global accelerator manager (GAM), shared L2 cache banks and shared network-on-chip (NoC) routers between multiple accelerators. All of the mentioned components are connected by the NoC. Accelerator nodes include a dedicated DMA-controller (DMA-C) and scratch-pad memory (SPM) for local storage and a small translation look-aside buffer (TLB) for virtual to physical address translation. GAM is introduced to handle accelerator sharing and arbitration. In this architecture we first propose a hardware resource management scheme, facilitated by GAM, for accelerator sharing. This scheme supports sharing and arbitration of multiple cores for a common set of accelerators, and it uses a hardware-based arbitration mechanism to provide feedback to cores to indicate the wait time before a particular resource becomes available. Second, we propose a lightweight interrupt system to reduce the OS overhead of handling interrupts which occur frequently in an accelerator-rich platform. Third, we propose architectural support that allows us to compose a larger virtual accelerator out of multiple smaller accelerators. We also implemented a complete simulation tool-chain to verify our ARC architecture. On a set of medical imaging applications (our initial application domain), ARC show significant performance improvement (on average 50 ×) and energy improvement (on average 379 ×) compared to an Intel Core i7 L5640 server running at 2.27 GHz.

Although ARC produces impressive performance and energy improvement, it has two limitations. One is that it has narrow workload coverage. For example, the highly specialized accelerator for denoise cannot be used for registration. The second limitation is that each accelerator has repeated resources, such as the DMA engine and SPM, which are underutilized when the accelerator is idle.

To overcome these limitations of ARC, we introduced CHARM: a composable heterogeneous accelerator-rich microprocessor design that provides scalability, flexibility, and design reuse in the space of accelerator-rich CMPs [Ref. 6]. We noticed that all the accelerators in the ARC for the medical imaging domain can be decomposed into a small set of computing blocks, such as 16-input polynomial, floating-point divide, inverse, and square root functions. These blocks are called the accelerator building blocks (ABBs). CHARM (shown in

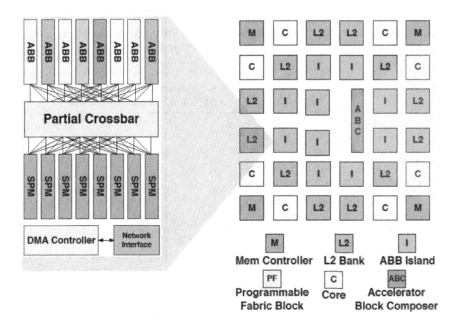

Fig. 1.2. Overview of the CHARM architecture.

Fig. 1.2) features a hardware structure called the accelerator block composer (ABC), which can dynamically compose a set of ABBs into a dedicated accelerator to provide orders-of-magnitude improvement in performance and power efficiency. Our compiler decomposes each compute-intensive kernel (candidate for accelerator) into a set of ABBs at the compilation time, and stores the data flow graph describing the composition. ABC uses this graph at runtime to compose as many accelerators as needed and available (based on free ABBs) for that kernel for acceleration. Therefore, although each composed accelerator is somewhat slower than the dedicated accelerator, we can potentially get many more copies of the same accelerator, leading to better acceleration results. Our ABC is also capable of providing load balancing among available compute resources to increase accelerator utilization. On the same set of medical imaging benchmarks, the experimental results on CHARM show better performance (almost 2 × better than ARC) and similar energy efficiency. Details are shown in Table 1.1.

Table 1.1.   Performance and energy-efficient comparison between multi-core CPU, GPU, FPGA, ARC and CHARM. Results are normalized to an Intel Core i7 (L5640 @ 2.27 GHz).

	GPU (NVIDIA Tesla M2075)	FPGA (Xilinx V6)	Monolithic Accelerators	Composable Accelerators
Deblur				
Performance	97 ×	25 ×	58 ×	107 ×
Energy	19 ×	130 ×	369 ×	261 ×
Denoise				
Performance	38 ×	12 ×	26 ×	37 ×
Energy	7.5 ×	89 ×	327 ×	308 ×
Segmentation				
Performance	52 ×	78 ×	79 ×	155 ×
Energy	2.4 ×	371 ×	201 ×	149 ×
Registration				
Performance	32 ×	24 ×	53 ×	109 ×
Energy	27.8 ×	31 ×	854 ×	1102 ×
Average				
Performance	50 ×	27 ×	50 ×	90 ×
Energy	10 ×	107 ×	379 ×	338 ×

More importantly, CHARM architecture provides better flexibility and wider workload coverage. By using the same set of ABBs designed for the medical imaging domain [Ref. 6], one can compose accelerators in other domains, such as navigation and vision, and still achieve impressive speedup and energy reduction.

The latest extension that we made to the CHARM architecture was finalized shortly after my visit to Imperial College London. Although CHARM has much better flexibility, it is possible that it misses some ABBs needed to compose some functions in a new application domain. To address this issue, we propose CAMEL: composable accelerator-rich microprocessor enhanced for longevity [Ref. 7]. CAMEL features programmable fabric (PF) to extend the use of ASIC composable accelerators to support algorithms that are beyond the scope of the baseline platform. Figure 1.3 shows the overall architecture diagram of CAMEL. Using a combination of hardware extensions and compiler support, we demonstrate an on average 11.6 × performance improvement and 13.9 × energy savings

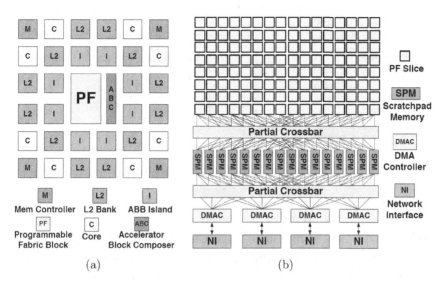

Fig. 1.3. (a) Overall architecture of CAMEL; (b) programmable logic used for ABB composition (from [Ref. 7]).

across benchmarks that deviate from the original domain for our baseline platform. More detail is available from [Ref. 7].

1.3. Memory Support for Accelerator-Rich Architectures

In most cases, the data access patterns of accelerators are known in advance. This is especially true with many image processing applications where accelerators typically process one tile of the image at a time. In this case, one uses buffers or SPM instead of cache to supply data to accelerators. However, since accelerators may not be used all the time, dedicated buffers can be wasteful. Therefore, we developed techniques to implement SPM in L1 or/and L2 caches. Here, I highlight two efforts in this direction.

The first effort is on adaptive hybrid L1 cache. The basic idea is to allow part of an L1 cache to be reconfigured as an SPM, as shown in Fig. 1.4. We add an extra bit to each cache block to indicate whether it belongs to cache or SPM, and add a mapping table that

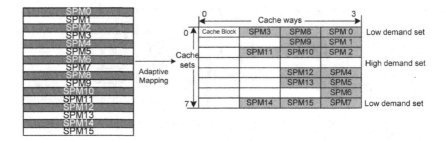

Fig. 1.4.   Customizable mapping of SPM in AH-cache (from [Ref. 8]).

maps the consecutive SPM blocks to a collection of cache blocks in different sets depending on the utilization of the sets (intuitively, underutilized sets contribute to blocks for the SPM). In fact, by reconfiguring part of the cache as software-managed SPM, hybrid caches can help handle both unknown and predictable memory access patterns to improve the cache efficiency for processor designs in general. To demonstrate this point, we improve the previous work on hybrid caches by introducing dynamic adaptation to the run-time cache behavior (dynamically remaps SPM blocks from high-demand cache sets to low-demand cache sets) [Ref. 8]. This approach achieves 19%, 25%, 18% and 18% energy-runtime-production reductions over four previous representative cache set balancing techniques on a wide range of benchmarks (previous cache set balancing techniques are either energy-inefficient or require serial tag and data array access). We leveraged domain-specific knowledge to allocate blocks in the SPM, examining data re-use patterns in applications to maximize the efficiency of the SPM.

The hybrid L1 cache is helpful for the case with one (or few) accelerators tightly coupled with a processor core so that it may share its L1 cache as a buffer. In the case of an accelerator-rich CMP, we need an efficient way to share the buffers among different accelerators, and share the accelerator buffers with the processor cache. Existing solutions allow the accelerators to share a common pool of buffers (accelerator store [Ref. 9]) or/and allocate buffers in cache (BiC [Ref. 10]). In order to achieve higher flexibility and better efficiency, we introduced a buffer-in-NUCA (BiN) [Ref. 11] scheme with the following features: (1) a dynamic interval-based

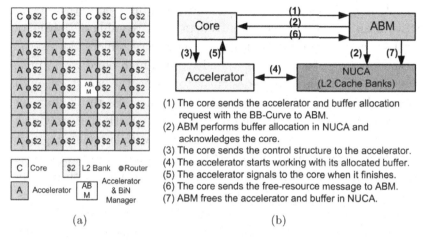

(a) (b)

Fig. 1.5. (a) Overall architecture of ARC with BiN; (b) communication between core, ABM, and accelerators (from [Ref. 11]).

global buffer allocation method to assign shared buffer spaces to accelerators that can best utilize the additional buffer space, and (2) a flexible and low-overhead paged buffer allocation method to limit the impact of buffer fragmentation in a shared buffer, especially when allocating buffers in a non-uniform cache architecture (NUCA) with distributed cache banks. BiN was implemented on top of the accelerator-rich CMP ARC. It has a global accelerator buffer manager (ABM), which works with the global accelerator manager to allocate buffers (including both the buffer size and locations) at the time of accelerator allocation. The overall ARC architecture with BiN and the communication flow is shown in Fig. 1.5. Experimental results show that when compared to accelerator store and BiC, BiN improves performance by 32% and 35% and reduces energy by 12% and 29%, respectively. More detail is available in [Ref. 11].

1.4. Network-on-Chip (NoC) Support of Accelerator-Rich Architectures

In accelerator-rich architectures, on-chip interconnects need to provide high bandwidth between the accelerators and buffers (e.g.

allocated in L2 cache by BiN in the preceding section), and also between buffers and memory controllers (for streaming data from off-chip) for an extended period of time. It is very difficult for existing packet-switching based NoCs to deliver such high bandwidth, unless they are significantly over-designed to cover the worst case, but this results in large area and power overhead. We think that such requirements can be addressed from our early work that provided a combination of packet switching and circuit switching via integration of radio frequency interconnect (RF-I) through on-chip transmission lines overlaid on top of traditional NoC implemented with RC wires [Ref. 12]. The RF-I goes to all the tiles on-chip, and each tile has a RF-I transmitter and receiver. By tuning the transmitter and receiver of two different tiles to the same carry frequency, one can create a dedicated link (or shortcut) between these two tiles, i.e. a customized interconnect. Research in [Ref. 12] showed that in addition to the latency advantage of RF-I (signals are transmitted at the speed of light and can go across the chip in a single clock cycle), there are three additional advantages of RF-I: 1) RF-I bandwidth can be flexibly allocated to provide an adaptive NoC; 2) RF-I can enable a dramatic power and area reduction by simplification of the baseline NoC (e.g. in terms of its link bandwidth), as the RF-I shortcuts can be customized to meet the peak communication demand; and 3) RF-I provides natural and efficient support for multicast. Based on these observations, a novel NoC design was proposed in [Ref. 12], exploiting dynamic RF-I bandwidth allocation to realize a reconfigurable NoC architecture. For example, the work in [Ref. 12] shows that using adaptive RF-I architecture on top of a mesh with 4B links only can match or even outperform the baseline NoC with 16B mesh links, but reduces NoC power by approximately 65%, including the overhead incurred for supporting RF-I. Figure 1.6 shows a mesh NoC with RF-I overlaid on top of it, and the customized shortcuts (based on a given communication pattern). We are interested in using this kind of hybrid NoCs with a mix of packet switching and circuit switching (implemented by the RF-I shortcuts) to provide high bandwidth between the accelerators and buffers (e.g. allocated in the L2 cache by BiN in the preceding section), and also between buffers and memory

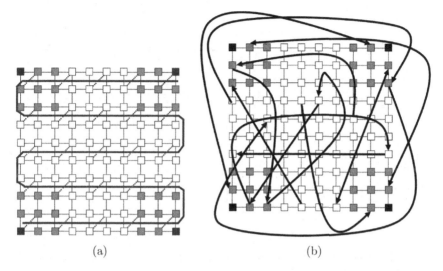

Fig. 1.6.   (a) RF-I overlaid with a mesh NoC; (b) RF-I shortcuts for customized communication patterns (from [Ref. 12]).

controllers (for streaming data from off-chip). This is an area of ongoing research.

1.5. Prototyping of an Accelerator-Rich Architecture

During my talk at Imperial College London, I mentioned our ongoing effort to prototype an accelerator-rich architecture in a large FPGA. I am glad to report that we made good progress in this area. We prototyped a general framework of an accelerator-rich architecture, named PARC, on a multi-million gate FPGA (Xilinx Virtex-6 FPGA) [Ref. 13]. We developed a set of system IPs that serve a sea of accelerators, including an IO memory management unit (IOMMU), a GAM, and a dedicated interconnect between accelerators and memories to enable memory sharing [Ref. 14]. Figures 1.7 and 1.8 show the system-level diagram and the floor plan implementation of PARC with four accelerators (acc gradient, acc Rician, acc Gaussian, and acc seg). Table 1.2 shows the performance and energy gain over the embedded MicroBlaze core on the Xilinx Virtex-6 FPGA and the embedded ARM Cortex A9 core in the Xilinx Zynq FPGA. It

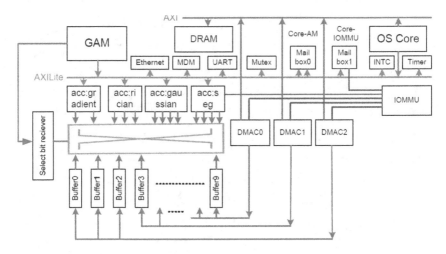

Fig. 1.7. Overview of our PARC implemented in the Xilinx Virtex-6 XC6V
LX240T FPGA. GAM = global accelerator manager (from [Ref. 13]).

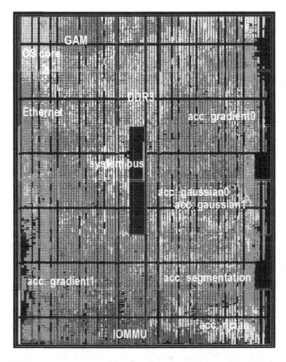

Fig. 1.8. Floorplan of PARC implemented in the Xilinx Virtex-6 FPGA (from
[Ref. 13]).

Table 1.2. Speed-up and energy savings of the accelerators in PARC (from [Ref. 13]).

	Segmentation	Gaussian	Gradiant + Racian
Xilinx Microblaze Processor in FPGA @ 100 MHz			
runtime (s)	1.4e3	1.7e2	3.3e3
energy (J)	546	66.3	129
Dual-Core ARM Cortex-A9 MPCore @ 800 MHz			
runtime (s)	0.597 (1×)	0.301 (1×)	0.862 (1×)
energy (J)	0.299 (1×)	0.150 (1×)	0.431 (1×)
Accelerator in Our System in FPGA @ 100 MHz			
runtime (s)	0.056 (10.7×)	0.066 (4.6×)	0.060 (14.4×)
energy (J)	0.123 (2.4×)	0.145 (1.0×)	0.132 (3.3×)
8-core Xeon Server E5405 @ 2 GHz			
runtime (s)	0.405 (1×)	0.109 (1×)	0.106 (1×)
energy (J)	4.056 (1×)	1.064 (1×)	0.992 (1×)
Accelerator in Our System projected on 45 nm ASIC			
runtime (s)	0.014 (28×)	0.016 (6.6×)	0.015 (7.1×)
energy (J)	0.010 (395×)	0.012 (87×)	0.011 (90×)

also shows the projected performance and energy of PARC if it is implemented in a 45 nm ASIC and the comparison with an 8-core Intel Xeon server.

1.6. Software Support

Software support is crucial for the successful design and deployment of the accelerator-rich architectures. We can classify the software support into two categories, one for the creation of accelerator-rich architectures and another for mapping and programming on an accelerator-rich architecture.

Software support for accelerator-rich architecture creation:

- **Accelerator synthesis:** We synthesize all accelerators automatically from high-level C/C++ specification to Verilog or VHDL designs using AutoPilot [Ref. 15] from AutoESL (based on xPilot [Ref. 16] from UCLA); this supported C-to-RTL designs for both ASIC and FPGA designs. After AutoESL was acquired by Xilinx, AutoPilot was integrated into the Xilinx Vivado design tool suite and renamed Vivado HLS. However, it supports Xilinx FPGA designs only. Other C-to-RTL tools, such as C-to-Silicon from Cadence [Ref. 17], Catapult C from Calypto [Ref. 18], and Cynthesizer from Forte [Ref. 19], can also be used for accelerator synthesis and generation on ASICs.

- **Accelerator virtualization:** Furthermore, we need tools to provide accelerator virtualization — that is, using available physical accelerators to implement more complex accelerators of the same or similar types (e.g. implementing a 1024-point FFT function using a 128-point FFT accelerator).

- **Accelerator identification:** Currently, we identify accelerators manually through profiling. In the future, we plan to develop tools to automatically identify accelerator candidates. One direction is to use pattern-mining techniques [Refs. 20 and 21] on the control and data-flow graph (CDFG), in a way similar to how we generate the customized vector instructions [Ref. 22].

- **Platform synthesis:** Together with the IPs we developed in the PARC prototype, we also developed an automated flow with a number of IP templates and customizable interfaces to a C-based synthesis flow to enable automated generation of an accelerator-rich platform on a Xilinx FPGA, so that one can rapidly design and update accelerator-rich architectures. We achieve significant productivity improvement with such an automated flow. For example, to add ten new accelerators of a certain type, we only need to add three lines of code in our configuration file, which will generate over 3,000 lines of C and RTL code in PARC [Ref. 12].

Mapping and programming on an accelerator-rich architecture

There are many demands for efficient mapping and optimization tools to map an application to an architecture-rich platform with all the customization capabilities as we described earlier in this paper. Here are some examples.

- We need efficient ways to map a large computation kernel into a set of ABBs in a composable accelerator-rich architecture, as in CHARM, with consideration of area, communication, and load-balancing optimization.
- We need compilation support to efficiently use the hybrid L1 cache with SPMs. We made some progress in this area [Ref. 23], but more studies are needed.
- We need tools to synthesize customizable NoCs with RF-Is to decide how to dynamically add shortcuts and construct the route. For example, our work in [Ref. 24] shows that routing in an irregular NoC (a result of adding a RF-bus to the underlying mesh-based NoC) may deadlock, and we developed efficient NoC construction and routing algorithms to avoid the deadlock.

1.7. Concluding Remarks

Driven by the need for higher energy efficiency, we believe that future computing platforms will be heterogeneous, customizable, and rich in accelerators, including both dedicated accelerators and composable accelerators from either accelerator building blocks or fine-grain programmable fabrics. In fact, such accelerator-rich platforms will happen at different levels: chip-level, rack-level, and data-center level. This offers many research opportunities for architecture, compiler, and runtime system support. I hope that what I have presented here will engender an increased interest from the research community, and encourage a focus on accelerator-rich computing platforms.

Acknowledgements

The CDSC is funded by the NSF Expeditions in Computing program (award CCF-0926127). Most of the advancements summarized in this paper are the joint work of CDSC faculty and students — especially those who worked on the design, prototyping, and software support for the accelerator-rich architectures: Yu-ting Chen, Mohammad Ali Ghodrat, Michael Gill, Beayna Grigorian, Hui Huang, Chunyue Liu, Glenn Reinman, Bingjun Xiao, Bo Yuan, and Yi Zou. The complete list of all CDSC faculty and students is available at www.cdsc.ucla.edu. The author would also like to thank the support of the Distinguished Visiting Fellow Program from the Royal Academy of Engineering (UK) and the warm hospitality of Professors Wayne Luk and Peter Cheung that the author received during his visit to Imperial College London in July 2012.

References

1. P. Schaumont and I. Verbauwhede. Domain-specific Codesign for Embedded Security, *Computer*, 36(4), 68–74, 2003.
2. G. Venkatesh *et al.* Conservation Cores: Reducing the Energy of Mature Computations, *ACM SIGARCH Computer Architecture News*, 38(1), 205–218, 2010.
3. H. Esmaeilzadeh *et al.* Dark Silicon and the End of Multicore Scaling, in *Proc. International Symposium on Computer Architecture*, pp. 365–376, 2011.
4. J. Cong *et al.* Customizable Domain-specific Computing, *IEEE Design & Test of Computers*, 28(2), 6–15, 2011.
5. J. Cong *et al.* Architecture Support for Accelerator-rich CMPs, in *Proc. Design Automation Conference*, pp. 843–849, 2012.
6. J. Cong *et al.* CHARM: A Composable Heterogeneous Accelerator-rich Microprocessor, in *Proc. International Symposium on Low Power Electronics and Design*, pp. 379–384, 2012.
7. J. Cong *et al.* Composable Accelerator-rich Microprocessor Enhanced for Adaptivity and Longevity, in *Proc. International Symposium on Low Power Electronics and Design*, pp. 305–310, 2013.
8. J. Cong *et al.* An Energy-efficient Adaptive Hybrid Cache, in *Proc. International Symposium on Low Power Electronics and Design*, pp. 67–72, 2011.
9. M. Lyons *et al.* The Accelerator Store framework for High-performance, Low-power Accelerator-based Systems, *Computer Architecture Letters*, 9(2), 53–56, 2010.

10. C.F. Fajardo *et al.* Buffer-Integrated-Cache: A Cost-effective SRAM Architecture for Handheld and Embedded Platforms, in *Proc. Design Automation Conference*, pp. 966–971, 2011.
11. J. Cong *et al.* BiN: A buffer-in-NUCA Scheme for Accelerator-rich CMPs, in *Proc. International Symposium on Low-power Electronics and Design*, pp. 225–230, 2012.
12. M.C.F. Chang *et al.* Power Reduction of CMP Communication Networks via RF-Interconnects, in *Proc. International Symposium on Microarchitecture*, pp. 376–387, 2008.
13. Y.-T. Chen *et al.* Accelerator-Rich CMPs: From Concept to Real Hardware, in *Proc. International Conference on Computer Design*, pp. 169–176, 2013.
14. J. Cong and B. Xiao. Optimization of Interconnects Between Accelerators and Shared Memories in the Dark Silicon Age, in *Proc. International Conference on Computer-Aided Design*, pp. 630–637, 2013.
15. J. Cong *et al.* High-level synthesis for FPGAs: From prototyping to deployment, *IEEE Transactions on Computer-Aided Design of Integrated Circuits and Systems*, 30(4), 473–491, 2011.
16. J. Cong *et al.* Platform-based Behavior-level and System-level Synthesis, in *Proc. International SOC Conference*, pp. 199–202, 2006.
17. Cadence C-to-Silicon Compiler. [Online] Available at: http://www.cadence.com/products/sd/silicon_compiler/[Accessed 20 April 2014].
18. T. Bollaert. "Catapult synthesis: a Practical Introduction to Interactive C Synthesis", in eds. P. Coussy and A. Morawiec, *High-Level Synthesis*, pp. 29–52. Springer, Netherlands, 2008.
19. M. Meredith. "High-level SystemC Synthesis with Forte's Cynthesizer", in eds. P. Coussy and A. Morawiec, *High-Level Synthesis*, pp. 75–97. Springer, Netherlands, 2008.
20. J. Cong and W. Jiang. Pattern-based Behavior Synthesis for FPGA Resource Reduction, in *Proc. International Symposium on Field Programmable Gate Arrays*, pp. 107–116, 2008.
21. J. Cong, H. Huang, and W. Jiang. A Generalized Control-flow Aware Pattern Recognition Algorithm for Behavioral Synthesis in *Proc. Design, Automation & Test in Europe Conference & Exhibition*, pp. 1255–1260, 2010.
22. J. Cong *et al.* Compilation and Architecture Support for Customized Vector Instruction Extension, in *Proc. Asia and South Pacific Design Automation Conference*, pp. 652–657, 2012.
23. J. Cong *et al.* A Reuse-aware Prefetching Scheme for Scratchpad Memory, in *Proc. Design Automation Conference*, pp. 960–965, 2011.
24. J. Cong, C. Liu, and G. Reinman. ACES: Application-specific Cycle Elimination and Splitting for Deadlock-free Routing on Irregular Network-on-chip, in *Proc. Design Automation Conference*, pp. 443–448, 2010.

Chapter 2

Whither Reconfigurable Computing?

George A. Constantinides, Samuel Bayliss and David Boland

EEE Department, Imperial College London

We argue that FPGAs, more than two decades after they began to be used for computational purposes, have become one of the key hopes for extending the performance of computational systems in the era characterised by the end of Dennard scaling. We believe that programmability of future heterogeneous computing platforms has brought a new urgency to bear on several old problems in high-level synthesis for FPGAs. Our focus is on the two areas we believe are most underdeveloped in today's high-level synthesis software: effective utilisation of the numerical flexibility afforded by high-level correctness specifications, and application-specific memory subsystem synthesis. We conclude with our perspective on the likely future evolution of the field.

2.1. A Selective Context

We present below a necessarily rather narrow view of the evolution of the FPGA and the microprocessor, highlighting the interaction between the two and the major external drivers.

The field-programmable gate array (FPGA) was invented in the 1980s, but it developed and matured in the 1990s. Already in the early 1990s, academic conferences started to appear that were largely dedicated to the potential these devices had to implement computation, such as the first FPL, held in 1991 in Oxford. However, the nature of such devices has transformed over recent years. Initially FPGAs were largely homogeneous architectures, consisting of a large number of very fine grain logic cells. Responding to the nature of the new application areas, manufacturers evolved the FPGA architecture, first incorporating larger RAM blocks [Ref. 1] and then dedicated multiplication logic [Ref. 2]. Today, modern FPGA architectures are highly heterogeneous devices, containing logic cells, embedded RAM and DSP functionality, high speed transceiver circuitry and microprocessors. It is worth noting that the majority of these components, present as hard IP within an FPGA, could be implemented using lookup table functionality. However, to do so would be either too large, too slow, or consume too much power to be worthwhile [Ref. 3]. Thus through one lens we can see the evolution of the FPGA in recent years as a conscious decision to move away from having all area devoted to simple fine-grain units and their fine grain interconnect, towards the *specialisation of circuitry* to perform certain common tasks or classes of tasks.

In a sense, the evolution of the general purpose processor has mirrored the evolution of the FPGA over the same timeframe. Traditional, latency-driven, computer architecture largely consisted of utilising all available silicon in order to keep a single (or small number of) computational unit as busy as possible. This resulted in a very large amount of silicon and power consumption devoted to caching in particular, as well as various micro-architectural innovations to avoid latency-consuming pipeline stalls [Ref. 4]. These processors have formed the core of general purpose computer design for several decades. The most significant innovation to arise as a result has been the GPGPU, which has delivered major performance improvements in certain domains by explicitly abandoning some of the received wisdom of computer architecture, a process referred to by Bill Dally as 'the end of denial architecture' [Ref. 5]. GPGPU computing achieves its performance by using an explicitly software-managed memory hierarchy, returning hardware to computation, and using an abundance of threads to hide pipeline stalls. We may therefore view the evolution of the microprocessor as a conscious decision to move away from one complex unit towards dedicating areas to a large number of much simpler units and their interconnect; in a certain sense this is a mirror of the evolution of the FPGA.

It is no accident that the co-evolution of FPGA and microprocessors is now reaching the point of blurred boundaries. On the
microprocessor side of the picture, this has largely been driven by the
recent failure of Dennard scaling. Dennard scaling [Ref. 6] provided
a road map of how to scale various parameters under manufacturer
control, such as supply voltage, in response to the geometric scaling
of VLSI given by each processor generation. The recent deviation
from Dennard scaling, largely driven by power consumption concerns
[Ref. 7] has forced the general purpose processor industry to look
beyond clock frequency as the driver for performance. While high performance for throughput-dominated applications with embarrassing
levels of parallelism can be achieved in a direct way using the GPGPU
approach, latency constraints and algorithm bottlenecks mandate a
more heterogeneous approach [Ref. 8]. On the FPGA side of the
picture, silicon and power inefficiencies combined with new market
opportunities have driven the evolution of FPGA architecture to its
present heterogeneous state.

The future of manycore computing using traditional — but
simple — microprocessor cores is not rosy. In a landmark paper,
Berger *et al.* [Ref. 9] make a detailed study of the power/area
and power/performance tradeoffs available across a spectrum of
processor designs. Using predictions of future technologies, even with
extremely parallel workloads, the performance to be gained by using
more, simpler, traditional processors, is bounded from above by
factor of only about 4–8× over the next decade, far slower than
the historical trends. This is largely due to power consumption
limitations, resulting in the spectre of *dark silicon*, transistors that
may be present on a device but cannot be powered on simultaneously
without overloading the power limitations. The conclusion is clear:
to move beyond such limits, it becomes necessary to improve the
Pareto tradeoff itself, rather than simply move towards more, simpler,
processors on the Pareto front. The only clear way to do this is
through *circuit specialisation*: creating parts of a processor that are
specialised to particular commonly occurring tasks, and avoiding
the energy inefficiency in using general purpose architectures for

these tasks. This is exactly the area where the reconfigurable computing community has a head start, and can provide direction to the general purpose architecture community.

It seems inevitable that future computer architecture will therefore be programmable, contain elements of application-specific or domain-specific architecture, and be highly heterogeneous in nature. The major challenge is how to efficiently and effectively compile applications onto such platforms. This is a challenge that must be overcome, but one that is no longer faced by the FPGA community alone, as was often the case in the early efforts of high-level synthesis for reconfigurable computing. The coming industrial turn towards heterogeneous parallel computing opens many doors.

2.2. The Promise and the Challenges

High level synthesis for reconfigurable computing has made great strides recently. The Autopilot tool [Ref. 10] now included within the Vivado design suite is a high-quality high-level synthesis environment, using C as the input language. Academic efforts such as LegUp [Ref. 11] also point to a promising future for FPGA-based high level design. However, existing solutions for high-level synthesis do not — in our opinion — adequately address memory systems. It should not be up to the programmer to explicitly manage the transfer of data between external memory of various types (SDRAM, SRAM, etc.) and on-chip memory. Equally, we should not squander the potential of FPGA architectures by applying general purpose microprocessor cache schemes within an FPGA. In our view, it is high time that tools for customisation of computational circuitry were matched by aggressive tools for customisation of memory subsystem design. By pushing the complexity into the synthesis tool, we believe that significant performance advantages can be obtained without the area or energy overhead of caching schemes. This is the topic we consider in Sec. 2.4. We note that the degree of predictability of memory accesses largely defines the potential improvement possible by a customised memory system and that often memory accesses are very predictable in nature, especially for embedded applications, which we believe form the key driver for next-generation computer architecture.

The other area that is poorly covered by existing high-level design flows is the automation of the selection of numerical representation and precision. The designer of any hardware accelerator for a numerically intensive algorithm knows that this is one of the areas where customised logic can result in huge performance gains, and will naturally ask 'should I use floating-point, fixed-point, or some more esoteric number system to perform this task', 'how precise do my internal results really need to be', etc. These questions remain largely unautomated. As a result, designers will again often ape the systems used in general purpose processor designs, such as IEEE standard floating-point arithmetic as the 'gold standard' of real number representation. There are two problems with this approach. Firstly, it does not work: typically a designer will want to perform operations in a different order to that expressed in the original code, in order to improve hardware efficiency, for example by applying the associative law to regroup addition into a tree structure: $a+(b+(c+d)) = (a+b)+(c+d)$, a law that holds for real numbers *but does not hold for floating-point* thus raising questions of correctness. Usually, whether formalised or not, there will be some notion of an acceptable numerical result, which can be used to drive such decisions; indeed, without such a notion, it becomes impossible to demonstrate that the behaviour of even the original source code is acceptable. We strongly advocate the formalisation of such specifications. This leads us to the second problem: the designer operates with 'one hand tied behind her back' by being forced to replicate the hardware structures present in general purpose processors, which may be grossly inefficient for the problem at hand. Once a formal specification of numerical correctness is available, the designer, and the synthesis tool, should be free to produce any hardware structure meeting that specification, playing to the advantages of the underlying architecture. Thus the same algorithmic specification may map automatically to a mixed-precision implementation in a GPU, a double precision implementation in a CPU, and a fixed-point implementation in an FPGA. No two of these implementations may produce the same bit pattern at their outputs, but all should be verifiable with respect to the formal correctness criteria. This is the topic we consider in Sec. 2.3. We note that, while such freedom can be exploited in all numerical applications, the degree of freedom is particularly great in embedded applications, where specifications of correctness tend to be expressible at very high levels of abstraction, leaving lots of freedom for an advanced design tool to explore, e.g. a controller for an aircraft might mandate stability and minimisation of fuel consumption of the aircraft [Ref. 12]; a much higher level of abstraction than bit-level equivalence to a golden C model!

2.3. Numerical Behaviour

When creating digital hardware architectures, one must first select a finite precision number system to represent numerical data. Since this number system can only represent a subset of real numbers, rounding will often occur after an arithmetic operation so as to represent values using the chosen number system. Whilst the error introduced by the rounding of any single value may be small, over the course of an algorithm the accumulation of these errors can cause a significant deviation from the desired result.

A simple tactic to minimise this error would be to err on the side of safety and select a number system that has much greater precision than necessary to obtain the desired quality of output, if such a precision can be determined. However, this will come at a substantial cost in terms of performance. As an example, recent figures for the difference in performance, in terms of peak theoretical FLOPs, between single and double precision is approximately a factor of 2 to 3 for a CPU [Ref. 13] or 24 for a GPU [Ref. 14].

Since arithmetic computation forms the heart of many high-performance digital systems, if we are to create efficient hardware accelerators, then we first need to select number systems with the minimum precision necessary to guarantee that our design criteria are met. Unlike CPUs and GPUs, FPGAs offer the freedom to fully customise the precision used throughout an accelerator. As a result, development of techniques to select an optimised number system have been an extensive research topic for the FPGA community over the past decade [Ref. 15, 16]. In this section, we first describe the state-of-the-art techniques that help us guarantee that a given number system for a hardware accelerator satisfies a numerical correctness criterion. We further discuss how these techniques can be enhanced so that they are applicable to a wide range of algorithms. Finally, we outline some of the future challenges for research in this field.

2.3.1. *Bounding numerical errors*

The most straightforward way to estimate the error of any hardware accelerator is through simulation; indeed, this is the main technique used by industry. Unfortunately, the size of the search space for the inputs will generally be too large to explore exhaustively; this means simulation may miss corner cases and under-allocate the number of bits for an accelerator. This is unacceptable in any safety-critical system, and in any case only works when there is a trusted, 'golden' reference model or method of certification available.

In contrast, analytical approaches provide guarantees that a design criterion will not be violated. Early analytical approaches were based on interval arithmetic (IA) [Ref. 17], affine arithmetic (AA) [Ref. 18] and LTI theory [Ref. 15]. Unfortunately, because IA and AA cannot find tight bounds on the worst case error, they will typically over-allocate bits for any nontrivial example. LTI theory is powerful enough to compute tight bounds, but it is restricted to the LTI domain and this does not include general multiplication, for example.

More recently, new approaches have been created which involve constructing polynomials to represent the worst-case range of intermediate variables throughout an algorithm. Through computing the lower (γ_{lower}) and upper (γ_{upper}) bounds of these polynomials, we can select a number system which prevents overflow. Furthermore, if we first construct a polynomial $\hat{p}$ representing the range of every intermediate variable in the presence of finite precision errors and a second polynomial p representing the range in infinite precision, then the extrema of the function $\frac{|p-\hat{p}|}{p}$ represent the worst-case relative error introduced by the use of finite precision arithmetic.

Table 2.1.   Construction of polynomials.

| x, y are inputs | |
| Δ is the error bound determined by the precision, so that $|\delta_i| \leq \Delta$ | |

Code	Polynomial Representation of Variable Value (Floating-Point)
$a = x^*y;$	$a = xy(1 + \delta_1)$
$b = a * a;$	$b = (xy(1 + \delta_1))^2(1 + \delta_2)$
$c = b - a;$	$c = [(xy(1 + \delta_1))^2(1 + \delta_2) - xy(1 + \delta_1)](1 + \delta_3)$

To create these polynomials, we use standard models to represent finite precision errors. When using fixed-point, provided there is no overflow, numerical errors are limited to one unit in the last place. If we choose an η-bit number system where the maximum value is 2^X, the worst-case rounding error for any fixed point number x is given by Eq. (2.1). It follows that the result of any scalar operation ($\odot \in \{+, -, *, /\}$) is bounded as in Eq. (2.2). Similarly, for floating-point, provided there is no overflow or underflow, for any real value x, the closest floating-point approximation $\hat{x}$ of x can be expressed as in Eq. (2.3), where η is the number of mantissa bits used. Once again, it follows that the floating-point result of any scalar operation ($\odot \in \{+, -, *, /\}$) is bounded as in Eq. (2.4).

Through applying these models of error to every computation in an algorithm, we can construct polynomials that represent the potential range of every intermediate variable. This is shown for a simple example in Table 2.1.

$$\hat{x} = x + \delta_1 \qquad (|\delta_1| \leq \Delta, \text{where } \Delta = 2^{X-\eta}). \qquad (2.1)$$

$$\widehat{x \odot y} = (x \odot y) + \delta_1. \qquad (2.2)$$

$$\hat{x} = x(1 + \delta_1) \qquad (|\delta_1| \leq \Delta, \text{where } \Delta = 2^{-\eta}). \qquad (2.3)$$

$$\widehat{x \odot y} = (x \odot y)(1 + \delta_1). \qquad (2.4)$$

While constructing these polynomials is straightforward, finding their extrema is computationally intractable. Instead, algorithms focus on finding a computationally tractable lower bound $\hat{\gamma}_{lower} \leq \gamma_{lower}$ and upper bound $\hat{\gamma}_{upper} \geq \gamma_{upper}$. Ideally we wish to find bounds such that $\gamma_{lower} - \hat{\gamma}_{lower}$ and $\hat{\gamma}_{upper} - \gamma_{upper}$ are as small as possible.

One of the latest and most powerful techniques to achieve this is based upon a result from real algebra discovered by Handelman [Ref. 19]. This states that a polynomial p is non-negative if and only if p has a *Handelman representation* of the form Eq. (2.5).

$$f = \sum_{\alpha \in \mathbb{N}^n} c_\alpha \prod_{i=1}^{m} g_i^{\alpha_i}, \qquad (2.5)$$

where each c_α is a positive constant, each g_i is a positive inequality and $\mathbb{N}$ is the set of natural numbers.

Using this result, we first re-write the bounds of a polynomial $\hat{\gamma}_{lower} \leq p \leq \hat{\gamma}_{upper}$ as two separate equations $\hat{\gamma}_{lower} - p \geq 0$ and $p - \hat{\gamma}_{upper} \geq 0$. If we can find a Handelman Representation to prove each inequality is non-negative, then we have found the lower and upper bounds of the polynomial. Heuristics which search for these representations have been shown to be able to compute much tighter bounds than IA or AA and enable us to create substantially smaller hardware [Ref. 20].

2.3.2. *Can we apply these techniques to general code?*

The techniques described in the previous section are powerful and have been shown to result in substantial performance improvements for some simple benchmarks. Unfortunately, the size of the polynomials can grow exponentially in the number of operations, meaning it would become too time consuming to be applicable to real benchmarks.

However, we can simplify large polynomials by replacing all terms that contribute little to the final result with a single term. Table 2.2 analyses a polynomial representing the range of a floating point addition of two variables. It calculates the worst-case range of every individual term in this polynomial. Clearly, several terms such as $100y_1\delta_1$, $100x_1\delta_1$ and $100x_1y_1\delta_1$ will have little impact on the final bounds. As such, if we replace them with a single new bounded variable, we shrink our polynomial with little impact on the final

Table 2.2. Potential contribution of each monomial in $(1 + x_1)$ $(1 + y_1)(1 + \delta_1)$.

Compute $a = x \bullet y$, where $x = [8; 12], y = [9; 11]$
in 6 bit floating-point
let $|x_1| \leq 0.2, |y_1| \leq 0.1, |\delta_i| \leq 2^{-6} \Rightarrow x \in 10(1 + x_1), y \in 10(1 + y_1)$
$a = 10(1 + x_1)10(1 + y_1)(1 + \delta_1)$
$\quad = (100 + 100x_1 + 100y_1 + 100\delta_1 + 100x_1y_1 + 100x_1\delta_1 + 100y_1\delta_1$
$\quad\quad + 100x_1y_1\delta_1)$

Term	Potential Contribution	Term	Potential Contribution
100	100	$100x_1$	±20
$100\delta_1$	±0.09765625	$100x_1\delta_1$	±0.01953125
$100y_1$	±10	$100x_1y_1$	± -2
$100y_1\delta_1$	±0.009765625	$100x_1y_1\delta_1$	±0.001953125

bounds. This simplification technique enables the earlier bounding procedure to be applied to much larger algorithms consisting of straight-line code [Ref. 21].

However, many algorithms cannot be converted into straight-line code; algorithms often contain complex control structures such as 'while' loops. The challenge with these structures is that finite precision errors may cause a 'while' loop to fail to terminate [Ref. 16]. Interestingly, these polynomial bounding procedures can also be useful in choosing sufficient precision to ensure that 'while' loops terminate.

One technique to prove program termination is based on the following steps:

(1) Construct a *ranking function* [Ref. 22], $f(x_1, \ldots, x_n)$ that maps every potential state within the loop to a positive real number.
(2) Prove that for all potential values of the variables $x_1, \ldots, x_n$ within the loop body, when the ranking function is applied to the loop variables before and after the loop transition statements, it always decreases by more than some fixed amount $\epsilon > 0$, *i.e.* $f(x'_1, \ldots, x'_n) \leq f(x_1, \ldots, x_n) - \epsilon$.

If we note that proving a ranking function decreases $(f(x'_1, \ldots, x'_n) \leq f(x_1, \ldots, x_n) - \epsilon)$ can be re-written as a question of non-negativity $(0 \leq f(x_1, \ldots, x_n) - f(x'_1, \ldots, x'_n) - \epsilon)$, then we can apply the same techniques that prove non-negativity to prove termination in finite precision arithmetic [Ref. 16].

2.3.3. *Next steps*

The techniques described in this section only touch the surface of research into automatically selecting the minimum precision necessary to meet design criteria. However, crucially they offer substantial progress in answering the following question: given a hardware architecture and word-length specification, will my design satisfy the specification? This will enable further research in this field; this includes delving deeper into techniques to assign the word-length for each individual operator in a large datapath and minimise the total area consumption [Ref. 23, 24], studying the links between how the order of operations in a hardware datapath can affect the error seen at the output and exploring the relationship between numerical precision and termination of iterative algorithms. Research into numerical behaviour has entered exciting times.

2.4. Memory Systems

In the preceding sections, we described how high-level specification of numerical accuracy can enable us to make more efficient use of silicon area. This in turn enables better performance where the number of parallel processing units can be increased, provided those units can be efficiently fed with data.

The most area-efficient technologies in common use for implementing commodity memory today (DRAM and Flash) have optimal process parameters that conflict with those needed to build fast logic. Wherever applications require large amounts of memory, that memory is implemented using a separate memory die. However, because device pin-density and off-chip switching frequencies have not scaled as rapidly as the exponential growth of transistors dedicated to logic datapath implementation, external memory bandwidth has increasingly become a performance bottleneck.

So it is critical to ensure that limited off-chip memory bandwidth is used efficiently. Herein lies a second challenge. DRAM memory structure is arranged in banks, rows and columns. Each row must be 'activated' before data held in columns within that row can be read or written. The row must then be 'precharged' before data in another row can be accessed. Timing parameters determined by physical DRAM memory array architecture constrain the minimum time between successive row activations. Over time, increasing memory clock frequencies mean that there is an ever larger penalty paid for random access to DRAM memory. In practical terms, this means there is a greater than 10× performance difference between the worst case and best case memory bandwidth obtained through different memory address sequences.

This has made it essential to develop memory subsystems which exploit the locality of memory accesses to provide the illusion of fast access to large amounts of memory. In a CPU, caches and dynamic memory controllers buffer and reorder memory requests to help ensure this happens. They typically must assume no prior knowledge of the sequence of memory requests from the datapath. Furthermore, CPUs implement non-deterministic bus interfaces which make memory performance difficult to analyse. Where a memory system is implemented in reconfigurable hardware, it can be customised for a specific application. Three key benefits can then be realised:

(1) fine grained on-chip memories provide a very large on-chip memory bandwidth to customised datapath,
(2) data buffered in those memories can be reused, reducing off-chip memory bandwidth requirements, and
(3) off-chip memory requests can be reordered to make the most efficient use of limited bandwidth.

The most memory intensive parts of a program tend to be in loops, so we target nested loops in our work [Ref. 25]. Static analysis to model the sequence of memory accesses which occur in nested loops can be done using the Polyhedral Model [Ref. 26,27]. From this analysis, automated tools allow us to synthesise a high performance application-specific memory system. In Sec. 2.4.1, we provide a brief overview of the Polyhedral Model. Section 2.4.2 describes a way in which this model can be used to build high performance application-specific memory systems.

2.4.1. *What is the Polyhedral Model?*

The Polyhedral Model represents a set of loop iterations as those integer vectors which satisfy a finite set of affine inequalities. The code in Fig. 2.1 shows a two level nested loop. The set of loop iterations is described by upper and lower bounds which are affine expressions of the surrounding loop variables (x_1 and x_2). The iterations of an n-level loop nest can be described implicitly as an integer set $\{\mathbf{x} \in \mathbb{Z}^n | A\mathbf{x} \leq \mathbf{b}\}$ where A is a $2n \times n$ integer matrix, $\mathbf{b}$ is a $2n$ integer column vector and the vector inequality is interpreted as $\mathbf{x} \leq \mathbf{y}$ iff $x_i \leq y_i$ for all i.

These iterations can be scheduled according to a linear mapping function which determines a partial ordering of those iterations. For the example given in Fig. 2.1, a mapping function $\sigma(\mathbf{x}) = (16\ 0)\left(\begin{smallmatrix} x_1 \\ x_2 \end{smallmatrix}\right)$ describes the iteration ordering. Each iteration $\mathbf{x}$ may access memory via memory accessing function(s) of the form $g_j(\mathbf{x}) = \mathbf{f}_j\mathbf{x} + h_j$.

Code that fits into this form is common in video processing and dense linear-algebra applications. Exact dependence analysis for code which can be described in this way is often tractable using integer linear programming techniques [Ref. 28]. The Polyhedral Model gives us a formal mathematical representation of the sequence of memory addresses accessed in the program. In Sec. 2.4.2, we show how

```
int t; int arr[256];
for (x1 = 0 ; x1 <= 15 ; x1++) {
    for (x2 = 0 ; x2 <= 15 ; x2++} {
        ... = function( arr [x1+ 16*x2] );
    }
}
```

Fig. 2.1. Example code for a two-level nested loop.

transformations applied to that formal representation can help build a high performance memory system.

2.4.2. Building high performance memory systems

In the preceding section, we showed how we could formally characterise the memory access requirements within a nested-loop structure. We can use this information to decouple the off-chip memory accesses from datapath logic using on-chip memory buffers. If we can transform code so that data is reused from the on-chip memory buffer, we can reduce the number of accesses to off-chip memory.

We can represent the specific 'row' and 'burst' accessed in each memory request by adding new dimensions to the loop-nest representation. If the size of each DRAM row is R words, the row accessed by memory address $\mathbf{fx} + h$ is given by $r = \mathbf{fx} + h$ div $R = \lfloor \mathbf{fx} + h/R \rfloor$ where $\lfloor \cdot \rfloor$ represents the $floor$ function. The columns within each row can be represented as non-overlapping bursts to take advantage of the multi-word burst accesses supported by modern memory devices. These can be represented by $u = \lfloor \mathbf{fx} - rR)/B \rfloor$ where a burst is B words long.

While neither of these is directly amenable to linear algebraic representation, we may note that from the properties of the floor function:

$$\left\lfloor \frac{\mathbf{fx} + h}{R} \right\rfloor - 1 < r \leq \left\lfloor \frac{\mathbf{fx} + h}{R} \right\rfloor, \qquad (2.6)$$

and

$$\left\lfloor \frac{\mathbf{fx} + h - rR}{B} \right\rfloor - 1 < u \leq \left\lfloor \frac{\mathbf{fx} + h - rR}{B} \right\rfloor. \qquad (2.7)$$

We can rewrite Eq. (2.6) and Eq. (2.7) as linear equalities as shown below in Eq. (2.8) and Eq. (2.9), without loss of information.

$$\mathbf{fx} + h - R + 1 \leq Rr \leq \mathbf{fx} + h \qquad (2.8)$$

$$\mathbf{fx} + h - rR - B + 1 \leq Bu \leq \mathbf{fx} + h - rR \qquad (2.9)$$

We can then add these four extra inequalities to those already present defining the loop bounds. This forms, for each memory reference, an augmented system of linear inequalities that completely capture not only the iteration space but also the specific SDRAM rows and bursts accessed within the innermost loop.

Using standard unimodular loop transformations, we can transform this augmented polyhedral representation to expose those occasions where data items are reused by multiple loop iterations. After transformation, those redundant dimensions which *only* represent data reuse can be projected out of the resulting polyhedral representation to produce code which fetches each memory item only once from off-chip memory. From this representation, we can use standard loop reordering transformations to move 'row' dimension to the outer-most level of the loop nest, improving data-locality.

Figures 2.2(a)–(d) illustrate this process for the example code shown in Fig. 2.1. The code in Fig. 2.2(a) is augmented with the row iterator 'r' and the burst iterator 'u' to form Fig. 2.2(b). The 'r' variable can then be hoisted to the outermost loop level to give Fig. 2.2(c). Note now that the x_2 variable is accessed only once in each loop iteration and can therefore be eliminated to give Fig. 2.2(d). The sequence of memory addresses accessed in Fig. 2.2(d) accesses the same rows and bursts of the original source code, but now does so in a more efficient order, since there are fewer row swaps (and their associated timing penalties) incurred.

When this technique is applied to code, it can significantly improve interface bandwidth efficiency. We show this in Fig. 2.3 for three benchmarks (Matrix–Matrix Multiply, Sobel Filter and Gaussian Backsubstitution) parameterised with reuse buffers inserted at different levels of the loop nest. The insertion of the buffer at the outermost level of the loop nest ($t = 1$) allows reordering of all memory accesses and means less than 10% of memory access cycles are spent idle whilst DRAM rows are swapped compared with >75% in the original code. The different levels of parameterisation allow a trade-off between performance and the amount of on-chip memory dedicated to data buffering.

```
char arr[256]; // off-chip memory
for (x₁ = 0; x₁ <= 15; x₁++)  {
  for (x₂ = 0; x₂ <= 15; x₂++) {
    .. = function( arr[x₁+16*x₂] );
  } }
```

(a) Original code

```
char arr[256]; // off-chip memory
for (x₁ = 0; x₁ <= 15; x₁++) {
  for (x₂ = 0; x₂ <= 15; x₂++) {
  // Note : / is integer division.
  // Note : r and u loops have one iteration.
    for ( r = (x₁+16*x₂)/16; r <= (x₁+16*x₂)/16; r++ ) {
      for ( u = (x₁+16*x₂-16*r)/4; \
                u <= (x₁+16*x₂-16*r)/4; u++ ) {
        ... = function( arr[x₁+16*x₂] )
      } }
  } }
```

(b) Augmented code

```
char arr[256]; // off-chip memory
char buff[16][4][4]; // on-chip memory
for (r = 0; r <= 15; r++)  {
  for (u = 0; u <= 3; u++) {
    buff[r][u][0..3] = burstread(r,u);
    for ( x₂=r; x₂<=r; x₂++) {
      for ( x₁=4*u; x₁<=4*u+3; x₁++){
        ... = function( buff[x₂][x₁/4][x₁%4] );
      } }
} }
```

(c) Intermediate code

```
char arr[256]; // off-chip memory
char buff[16][4][4]; // on-chip memory
for (r = 0; r <= 15 ; r++) {
  for (u = 0; u <= 3; u++) {
    buff[r][u][0..3] = burstread(r,u);
    for (x₁ = 4*u; x₁ <= 4*u+3; x₁++ ) {
      ... = function( buff[r][u][x₁-4*u] );
    } }
}
```

(d) Transformed code

Fig. 2.2.   Transformation steps in improving memory bandwidth efficiency.

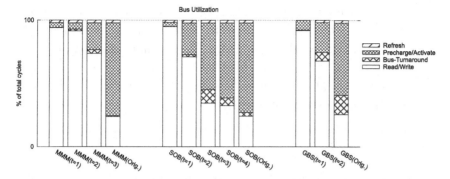

Fig. 2.3. Memory bandwidth improvements: Breakdown of interface commands by type.

2.4.3. *What might this enable us to do in the future?*

Looking beyond our existing work, the formal model of memory access provided by the Polyhedral Model is a promising representation for enabling other application-specific memory transformations. One emerging area of research is the exploration of how the Polyhedral Model enables the overlapping of off-chip memory operations with on-chip computation. This work makes use of mathematical advances [Ref. 29] which allow us to count the exact number of integer points contained with a polyhedron without enumerating them.

Knowledge of the exact lifetime of variables fetched into on-chip memory can enable more compact mapping of those variables into limited on-chip memory. Exploratory work on how to better utilise the multiple independent banks within a DRAM also seems like a promising direction, allowing us to further improve the efficiency of off-chip memory accesses.

Exact dependency analysis allows auto-parallelisation, but to support this, we need to ensure that enough on-chip memory ports are available to avoid contention. Emerging automatic array partitioning techniques [Ref. 30] ensure that contention for on-chip memory ports in minimised. This allows efficient use of the large on-chip bandwidth provided by block RAM resources which are ubiquitous in modern heterogeneous FPGAs.

The key theme is that there are significant opportunities opened up by expanding our synthesis tools to target complete reconfigurable systems including off-chip memory. The formal representation of memory access sequences provided by the Polyhedral Model allows tools to automatically produce efficient application-specific hardware with tailor-made memory systems.

2.5. Conclusion

Our view is that many problems in high-level design automation, once of concern to the small group of pioneers of reconfigurable computing, now arise in various guises in the much broader setting of computing generally, and embedded computing in particular. While high-level synthesis and compilation tools have progressed significantly over the past decade, we believe that there are two very significant gaps in existing tool flows: customisation of memory systems and auto-generation of finite precision arithmetic implementations. We have described our own approaches to these central problems. Our view is that the FPGA computing community is poised to play a central role in the evolution of computer architecture and compilers over the next decade. We must take up the baton.

2.6. Acknowledgements

We wish to acknowledge the inspirational mentorship of Professor Peter Y.K. Cheung, one of the small group of pioneers of FPGA-based computing. In addition, we would like to express our thanks to EPSRC for funding received to support the development of the ideas expressed in this paper (grants EP/G03157/1, EP/I012036/1, EP/I020357/1, EP/K034448/1).

References

1. S. Wilton. Architecture and Algorithms for Field-Programmable Gate Arrays with Embedded Memory, PhD thesis, University of Toronto, 1997.

2. S. Haynes, A.B. Ferrari, and P.Y.K. Cheung. Flexible Reconfigurable Multiplier Blocks Suitable for Enhancing the Architecture of FPGAs, in *Proc. Custom Integrated Circuit Conference*, pp. 16–19, 1999.

3. I. Kuon and J. Rose. Measuring the Gap between FPGAs and ASICs, *IEEE Transactions on Computer-Aided Design of Integrated Circuits and Systems*, 26(2), 203–215, 2007.

4. J.L. Hennessy and D.A. Patterson. *Computer Architecture- A Quantitative Approach (5 ed.)*, Morgan Kaufmann, Burlington, MA, USA, 2012.

5. W.J. Dally. The End of Denial Architecture and the Rise of Throughput Computing in *Proc. Design Automation Conference*, p. xv, 2009.

6. R. Dennard *et al.* Design of Ion-Implanted MOSFETs with Very Small Physical Dimensions, *IEEE Journal of Solid State Circuits*, SC-9(5), 256–268, 1974.

7. T.N. Mudge. Power: A First-Class Architectural Design Constraint, *IEEE Computer*, 34(4), 52–58, 2001.

8. A. Rafique, N. Kapre, and G.A. Constantinides. Avoiding Communication in GPU and FPGA-based Sparse Iterative Solvers: Algorithm and Architecture Interaction. In preparation.

9. H. Esmaeilzadeh *et al.* Dark Silicon and the End of Multicore Scaling, in *Proc. International Symposium on Computer Architecture*, 2011.

10. Z. Zhang *et al.* "AutoPilot: A Platform-Based ESL Synthesis System", in eds. P. Coussy and A. Morawiec, *High-Level Synthesis*, pp. 99–112, Springer, Netherlands, 2008.

11. A. Canis *et al.* LegUp: High-level Synthesis for FPGA-based Processor/Accelerator Systems, in *Proc. International Symposium on Field Programmable Gate Arrays*, pp. 33–36, 2011.

12. E. Hartley *et al.* Predictive Control of a Boeing 747 Aircraft using an FPGA, in *Proc. IFAC Nonlinear Model Predictive Control Conference*, pp. 80–85, 2012.

13. A. Vladimirov. Whitepaper: "Arithmetics on Intel's Sandy Bridge and Westmere CPUs: Not All FLOPs are Created Equal", 2012.

14. NVIDIA. NVIDIA Tesla Kepler GPU Computing Accelerators. [Online] Available at: http://www.nvidia.com/content/tesla/pdf/NV_DS_TeslaK_Family_May_2012_LR.pdf. [Accessed 23 April 2014].

15. G.A. Constantinides, P.Y.K. Cheung, and W. Luk. The Multiple Wordlength Paradigm, in *Proc. International Conference on Field-Programmable Custom Computing Machines*, 2001.

16. D. Boland and G. Constantinides. Word-length Optimization Beyond Straight Line Code, in *Proc. International Symposium on Field-Programmable Gate Arrays*, 2013.

17. R.E. Moore. *Interval Analysis*, Prentice-Hall, Englewood Cliff, NJ, USA, 1966.

18. L.H. de Figueiredo and J. Stolfi. Self-Validated Numerical Methods and Applications, *Brazilian Mathematics Colloquium Monographs*, IMPA/CNPq, Rio de Janeiro, Brazil, 1997.

19. D. Handelman. Representing Polynomials by Positive Linear Functions on Compact Convex Polyhedra, *Pac. J. Math.*, 132(1), 35–62, 1988.

20. D. Boland and G. Constantinides. Bounding Variable Values and Round-off Effects using Handelman Representations, *IEEE Transactions on Computer-Aided Design of Integrated Circuits and Systems*, 30(11), 1691–1704, 2011.

21. D. Boland and G.A. Constantinides. A Scalable Precision Analysis Framework, *IEEE Transactions on Multimedia*, 15(2), 242–256, 2013.

22. B. Cook, A. Podelski, and A. Rybalchenko. Proving Program Termination, *Communications of the ACM*, 2009.

23. D.-U. Lee *et al.* MiniBit: Bit-width Optimization via Affine Arithmetic, in *Proc. Design Automation Conference*, pp. 837–840, 2005.

24. D.M.H.-N. Nguyen and O. Sentieys. Novel Algorithms for Word-length Optimization, in *Proc. European Signal Processing Conference*, pp. 1944–1948, 2011.

25. S. Bayliss and G.A. Constantinides. Optimizing SDRAM Bandwidth for Custom FPGA Loop Accelerators, in *Proc. International Symposium on Field Programmable Gate Arrays*, pp. 195–204, 2012.

26. C. Lengauer. Loop Parallelization in the Polytope Model, in *Proc. International Conference on Concurrency Theory*, pp. 398–417, 1993.

27. W. Kelly and W. Pugh. *A Framework for Unifying Reordering Transformations, Technical Report UMIACS-TR-92-126.1*, University of Maryland, College Park, MD, USA, 1993.

28. W. Pugh. The Omega Test: A Fast and Practical Integer Programming Algorithm for Dependence Analysis, *Communications of the ACM*, 8, 4–13, 1992.

29. A. Barvinok. A Polynomial Time Algorithm for Counting Integral Points in Polyhedra when the Dimension is Fixed, in *Proc. Symposium on the Foundations of Computer Science*, pp. 566–572, 1993.

30. P. Li *et al.* Memory Partitioning and Scheduling Co-optimization in Behavioral Synthesis, in *Proc. International Conference on Computer-Aided Design*, pp. 488–495, 2012.

Chapter 3

An FPGA-Based Floating Point Unit
for Rounding Error Analysis

Michael Frechtling and Philip H.W. Leong

School of Electrical and Information Engineering,
The University of Sydney

Detection of floating-point rounding errors normally requires run-time analysis in order to be effective and software-based tools are seldom used due to the extremely high computational demands. In this chapter we present a field programmable gate array (FPGA) based floating-point co-processor which supports standard IEEE-754 arithmetic, user selectable precision and Monte Carlo arithmetic (MCA). This co-processor enables the detection of catastrophic cancellation and minimizing required floating-point precision in reconfigurable computing applications.

3.1. Introduction

IEEE-754 [Ref. 1] has long been the standard for computing using floating-point (FP) numbers; however, as a finite precision arithmetic system it is capable of anomalous results. Rounding error during computation can significantly reduce the accuracy of a computation, and result in errors many times larger than expected [Ref. 2]. In order to properly implement and verify numerical software, techniques to determine the effects of such errors are required. Monte Carlo arithmetic (MCA) [Ref. 3] can track rounding errors at run-time by forcing inputs and outputs to behave like random variables. Analysis of repeated operations turns an execution into trials of a Monte Carlo simulation allowing statistics on the effects of rounding errors to be obtained. MCA is typically performed using SW routines and as such its implementation involves a drastic reduction in performance. Field programmable gate arrays (FPGAs) offer a platform in which hardware (HW) acceleration can be applied to arbitrary algorithms.

In this work, we describe a complete HW accelerator for runtime error analysis. A novel MCA co-processor architecture is employed which is implemented entirely using standard floating-point cores. System-level performance measurements are described, and a comparison with an existing (SW) implementation made.

This work was influenced by Professor Cheung's seminal research on optimizing floating-point bit widths to advantageously utilize the reconfigurable nature of FPGAs. It is complementary to his approach to floating-point sensitivity analysis based on automatic differentiation [Ref. 4] and can be applied to customized hardware floating-point [Ref. 5] and dual-fixed point [Ref. 6] implementation schemes.

3.2. Background

3.2.1. *IEEE-754 floating point*

The binary IEEE-754 [Ref. 1] floating-point number system $\mathbb{F}(\beta, p, e_{min}, e_{max})$ is a subset of real numbers with elements of the form

$$x = (-1)^s m \beta^e. \tag{3.1}$$

The system is characterized by the radix β, which is assumed to be 2 in this paper, precision p, the exponent values $e_{min} \leq e \leq e_{max}$, the sign bit $s \in \{0, 1\}$ and the mantissa $m \in [0, \beta)$. Normalized values are most commonly used and are represented as a non-zero $x \in \mathbb{F}$ with $|m| \in [1, \beta)$ and $e_{min} < e < e_{max}$. De-normalized numbers are also supported and represent values of smaller magnitude than normalized numbers with $m \in [0, 1)$ and exponent $e = e_{min}$. Other classes of numbers including $+/-$ Zero, Infinity and Not a Number (NaN) are available with special formats. Without loss of generality, we assume the 32-bit IEEE single precision format in this paper with $p = 24$, $e_{min} = -125$, and $e_{max} = 128$. Real numbers are generally not exactly representable as FP numbers due to a number of factors including errors of measurement or estimation, quantization error or errors propagated from earlier parts of a computation. Although the IEEE-754 standard is used in all types of applications mostly without issue, if one or more non-exact numbers are subtracted, a loss of significant digits can occur due to normalization of the result [Ref. 2]. This phenomena is called *catastrophic cancellation* and is one of the major causes of loss of significance.

3.2.2. *Error analysis*

Several systems have been developed for performing run-time error detection and analysis. Interval Arithmetic (IA) represents a value x by an interval $[x_{lo}, x_{hi}]$. Intervals are propagated through the calculation $[a_{lo}, a_{hi}] - [b_{lo}, b_{hi}] = [a_{lo} - b_{hi}, a_{hi} - b_{lo}]$. IA can be used to track inexact values and rounding errors during computation; however, it often produces overly pessimistic error bounds [Ref. 7]. A limited number of hardware implementations of IA can be found in the literature [Refs. 8–10]. The CESTAC method [Ref. 11] is a special case of MCA that involves executing the same computation several times by randomly perturbing the rounding scheme of the arithmetic operators. By comparing the results from a number of different executions, the number of significant digits can be estimated. A hardware implementation of the CESTAC method has also been published [Ref. 12], but we are not aware of any system level implementation. An FPGA based implementation of MCA addition and multiplication with an area penalty of less than 22% over IEEE-754 was recently published by Yeung *et al.* [Ref. 13]. Compared with their implementation, this work presents a complete system for MCA rather than just an MCA core. In addition, while Yeung described a custom FPU for MCA, the FP computations in this work are performed using standard IEEE-754 FP primitives, providing significant benefits in terms of portability, flexibility and development time.

3.3. Monte Carlo Arithmetic

If x is a floating-point value of the form given in Eq. (3.1) we define the *inexact* function as

$$inexact(x, t, \xi) = x + \beta^{e_x - t}\xi \qquad (3.2)$$

$$= (-1)^{s_x}(m_x + \beta^{-t}\xi)\beta^{e_x}, \qquad (3.3)$$

where $x \in \mathbb{R}$; t is a positive integer representing the desired precision; ξ is a uniformly distributed random variable in the range $(-\frac{1}{2}, \frac{1}{2})$; $(\xi \in U(-\frac{1}{2}, \frac{1}{2}))$; and m_x, e_x are the mantissa and exponent of x. It is assumed that $1 < t \leq p$. An operation $\circ \in \{+, -, \times, \div\}$ is implemented as

$$x \circ y = round(inexact(inexact(x) \circ inexact(y))). \qquad (3.4)$$

Adjustments to input operands are referred to as *precision bounding* and are used to detect catastrophic cancellation, while adjustments to outputs are referred to as *random rounding* and are used to detect round-off error [Ref. 3]. The system developed for this paper performs precision bounding using multiple floating-point computations. Using this method the operation can be performed without modifying the internal architecture of the standard IEEE-754 FPU; however, the use of standard FP operations results in FP rounding being applied multiple times within a single MCA operation. In the case of traditional MCA the average of a Monte Carlo simulation can be used to estimate the true result of an operation sensitive to rounding error, and the relative standard deviation used to detect catastrophic cancellation. In the case of MCA implemented for this paper, the average value of the results of a Monte Carlo simulation cannot be used to estimate the true result of the tested operation, as this average is affected by the use of multiple rounding stages. In this case the system is only used for the detection of catastrophic cancellation. The value t shown in Eq. (3.4) is the *virtual precision* of the MCA operation. This value determines the number of places the value ξ is shifted to the right of the mantissa of the floating-point value x, and is used to control the level of random fluctuations applied during MCA. The virtual precision of the MCA operations is set to a positive integer less than or equal to the machine precision of the floating-point system being used:

$$1 \leq t \leq p. \tag{3.5}$$

A large t value will result in a smaller exponent value for the operand perturbation, increasing the accuracy of the operation. Similarly, a smaller t decreases the accuracy. In practice, variation of t is used to determine what effect lowering or increasing the precision has on the accuracy of an operation. The results of this analysis can then be used to determine an appropriate value for the machine precision p of a floating point operation that will maintain a required level of accuracy. The implementation developed for this paper performs *variable precision MCA*, and the value of t used by the co-processor can be modified at any time during execution. Further details are provided in Sec. 3.4.

3.4. System Implementation

The MCA FPU is an FPGA co-processor connected to a MicroBlaze soft processor through a AXI4-stream bus. The co-processor is capable of performing both standard floating-point arithmetic and MCA. The MCA FPU core was developed using the high-level C-to-RTL design software AutoESL (http://www.xilinx.com/tools/autoesl.htm). Using AutoESL, our MCA FPU is described using standard C statements and during synthesis and implementation, floating-point operations are translated into a set of floating-point modules based on the IEEE-754 floating-point library.

3.4.1. *MCA FPU implementation*

The MCA FPU is able to perform four basic arithmetic operations; add, subtract, multiply and divide. A fifth configure operation allows the precision value, t, and the MCA flag to be modified at run-time. The final operation combines the add and multiply operations to perform an FMA (fused multiply-add) operation, which calculates the result of the operation $r = (a * b) + c$. The arithmetic operations are implemented by coupling standard IEEE-754 floating-point operator primitives with a configuration register and a perturbation generation module. Each perturbation module is used to determine a value that will be added to the operation based on the value of the operands and a random number. Random numbers are generated using *Maximally Distributed Tausworthe Generators* (TRNG) [Ref. 14]. The configuration information consists of a 1-bit Boolean flag indicating MCA or IEEE-754 mode, and a 32-bit unsigned value for t. These are stored in the configuration register for access during subsequent operations. In IEEE mode, the FPU fully supports the standard. A description of how the operators are implemented is given below.

3.4.1.1. *MCA addition/subtraction*

Addition and subtraction operations are performed in terms of the *inexact* function (Eq. (3.3)) as follows:

$$x \pm y = round(inexact(x) \pm inexact(y)) \qquad (3.6)$$

$$= round((x + \xi_x) \pm (x + \xi_y)) \qquad (3.7)$$

$$= round(x \pm y + \xi), \qquad (3.8)$$

where $\xi_x, \xi_y \in U(-\frac{1}{2}, \frac{1}{2})$ and $\xi = \xi_x \beta^{e_x - t} + \xi_y \beta^{e_y - t}$. The magnitude of ξ can be calculated using only positive values via

$$|\xi| = \beta^{e_x - t}(|\xi_x| \pm |\xi_y|\beta^{-(e_x - e_y)}) \qquad (3.9)$$

and a equal probability choice of addition or subtraction is made. Note that $|\xi| \in \beta^{e_x - t}[0, 1)$ and its distribution depends on the value of x and y. The floating-point value ξ can be calculated by first using fixed-point arithmetic to produce separate values for the e_ξ and m_ξ then combining these values along with a randomly selected value for s_ξ in the correct format to produce a floating-point number:

$$m_\xi = \xi_x + [\xi_y \beta^{-(e_x - e_y)}], \ e_\xi = e_x - t. \qquad (3.10)$$

To perform this operation two random values for ξ_x and ξ_y must be calculated. These values will be used to form m_ξ and as such must be 24-bit normalized fixed-point values. Each value must also be in $U[0, \frac{1}{2})$. Two 32-bit values are produced (one from each TRNG) and the lower 22-bits assigned as the fixed-point value of ξ_x or ξ_y. The MSBs of each 32-bit number are used to calculate the sign bit s_ξ. Once the fixed point values of ξ_x and ξ_y have been produced the value of m_ξ can be calculated using fixed-point arithmetic At this point we have produced a value $m_\xi \in [0, 1)$. To produce the final value for ξ this value must be normalized, the $\beta^{e_x - t}$ shift applied and the value converted to IEEE-754 single precision format. This is done in the following stages:

(1) Determine the number of leading zeroes λ_ξ. The leading zero detector (LZD) used for the Monte Carlo FPU is based on the LZD found in [Ref. 15]. The mantissa value m_ξ is then shifted left or right depending on the value of λ_ξ forming the final 24 bit normalized mantissa.
(2) Calculate the 8-bit exponent value based on the value of e_x, λ_ξ and t.
(3) Merge the sign, exponent and mantissa values to form the single precision floating-point value using left-shifts to move the sign and exponent values to the correct location.

Once a value for ξ has been produced the second addition operation is performed, producing the final result.

3.4.1.2. *MCA multiplication*

The multiplication operation can be represented in terms of the *inexact* function shown in Eq. (3.4) as follows:

$$xy = round((x + \xi_x)(y + \xi_y)) \tag{3.11}$$
$$= round(xy + x\xi_y + y\xi_x + \xi_x\xi_y). \tag{3.12}$$

The perturbation values in the above equation can be expanded and simplified to the following:

$$\xi = \beta^{e_x + e_y - t}[m_x\xi_y + m_y\xi_x + \xi_x\xi_y\beta^{-t}]. \tag{3.13}$$

From the above equation it can be seen that the $\xi_x\xi_y$ term will be shifted to the right by $2t$ places during the operation. Calculation of the perturbation value including this term would require the precision of the FPU to be extended, either by modifying the internal architecture of the FPU core or by performing the Monte Carlo calculation in a higher precision format. This can be avoided by not including the $\xi_x\xi_y$ term in the calculation, which can be done without significantly affecting the results as the large right shift results in an extremely small value for $\xi_x\xi_y$ relative to $m_x\xi_y + m_y\xi_x$. The Monte Carlo multiplication operation can therefore be simplified to the following:

$$xy = round(xy + \xi), \qquad (3.14)$$

where $\xi_x, \xi_y \in U[0, \frac{1}{2})$ and $\xi = \beta^{e_x+e_y-t}[m_x\xi_y + m_y\xi_x]$. The magnitude of ξ can be calculated using only positive values via

$$|\xi| = \beta^{e_x+e_y-t} \left(|m_x\xi_y| \pm |m_y\xi_x| \right), \qquad (3.15)$$

where a randomized, equal probability choice of addition or subtraction is made. Note that $|\xi| \in \beta^{e_x+e_y-t}[0, 2)$ and its distribution depends on the values of x and y. In MCA multiplication a similar method to addition is employed to produce the perturbation value. Two TRNGs are first used to produce 24-bit fixed point random numbers and the corresponding sign bits. These represent $\xi_x, \xi_y \in (-\frac{1}{2}, \frac{1}{2})$. Using Eq. (3.15), each value is then multiplied by the mantissa of the relevant operand and the resulting values added together, resulting in a value for m_ξ. This process produces a value $m_\xi \in (-2, 2)$. This value is used to produce a single precision floating-point value for ξ as follows:

(1) Determine the number of leading zeroes λ_ξ in m_ξ and normalize.
(2) Calculate the exponent value e_ξ.
(3) Merge the sign, exponent and mantissa values to form the 32-bit floating-point perturbation value.

Once the final perturbation value ξ has been produced it is added to the initial result r' to produce the final result of the operation.

3.4.1.3. *MCA division*

The division operation differs from addition, subtraction and multiplication in that two individual floating-point perturbation values ξ_x and ξ_y are produced rather than a single perturbation value ξ. This operation can be described in terms of Eq. (3.4) as follows:

$$\frac{x}{y} = round\left(\frac{inexact(x)}{inexact(y)}\right) = round\left(\frac{x + \xi_x}{y + \xi_y}\right). \tag{3.16}$$

The above equation cannot be easily simplified to a point where a combined value for ξ can be calculated as for previously discussed operators. Separate perturbation values (ξ_x and ξ_y) are therefore calculated and applied to the x and y operands, requiring the precision of the division operation to be extended. This is done by performing single precision (32-bit) MCA division using double precision (64-bit) floating-point division. The perturbation values are calculated as follows:

$$\xi_x = \beta^{e_x - t} U\left(-\frac{1}{2}, \frac{1}{2}\right), \quad \xi_y = \beta^{e_y - t} U\left(-\frac{1}{2}, \frac{1}{2}\right). \tag{3.17}$$

Each perturbation value is applied to the relevant operand using a standard IEEE-754 floating-point addition operation, after which a standard IEEE-754 double precision division operation is performed. Although this calculation requires a total of three FP operations, calculation and addition of the two perturbation values can be performed in parallel, and the only increase in overhead over addition/ multiplication is from the double precision division operation. MCA division is performed as follows. Two TRNGs are used to produce 24-bit fixed point mantissa values for ξ_x and ξ_y and their corresponding sign bits. The values are then converted to single precision floating-point format:

(1) Determine the number of leading zeroes λ_{xi} in each mantissa and normalize.
(2) Calculate the exponent values.
(3) Merge the sign, exponent and mantissa values.

Once the perturbation values ξ_x and ξ_y have been calculated the final result is calculated.

3.5. Testing Methods

Testing of the FPU was conducted by comparing the performance of the co-processor to a SW implementation of MCA. The test routines used are based on routines used by Parker in [Ref. 16], downloadable from http://www.cs.ucla.edu/~/stott/mca. Details of equipment and parameters are in Table 3.1. In order to compile unmodified C source code to use MCA, two different versions of the gcc software floating-point library were developed. For the FPGA case, a library in which the addsf3, subsf3, mulsf3 and divsf3 were changed to utilize the MCA co-processor was created. FP operations can then be redirected to the appropriate subroutine by invoking gcc with the -msoft-float option. PC implementations of MCA were compiled in a similar fashion using a different, software-only MCA library. For each test case discussed below three tests were performed. The first was on a PC using a floating-point unit (SW-FP); the second a PC with software MCA (SW-MCA); and finally, an FPGA using the Monte Carlo co-processor (HW-MCA).

Table 3.1.   System parameters.

Item	Version/Description
FPGA Parameters	
ISE Version	13.2
FPGA	Virtex-6 LX240T (Speed Grade 3)
FPGA Board	Xilinx ML-605 Development Board
Processor Clock Speed	150 MHz
MCA Core Clock Speed	150 MHz
PC Parameters	
CPU	Intel Core 2 Duo 3 GHz
Memory	4 GB
OS	Ubuntu 12.04 32-bit
GCC Version	4.7.0

3.5.1. Cancellation (Knuth) test

The cancellation test performs a simple associativity test by calculating

$$u = (x + y) + z, \quad v = x + (y + z). \tag{3.18}$$

Over the real numbers, u should equal v; however, for the values $x = 11111113.0$, $y = -11111111.0$, $z = 7.5111111$ catastrophic cancellation occurs. Using these values the difference between $|u|$ and $|v|$ is calculated over 1000 samples and the standard deviation of the results is used to determine the accuracy of the calculation.

3.5.2. Cosine test

The cosine test calculates the cosine function using a power series expansion:

$$\cos(z) = 1 - \frac{1}{2!}z^2 + \frac{1}{4!}z^4 - \frac{1}{6!}z^6 + \frac{1}{8!}z^8 - \cdots \tag{3.19}$$

for $z \in [0, \pi]$. For each value of z over n steps a set of 100 samples are calculated and compared to the value of a single precision FP calculation of the same value, and the accuracy of the calculation measured at each step.

3.5.3. Kahan test

The Kahan test performs an evaluation of a rational polynomial

$$rp(x) = \frac{622 - x(751 - x(324 - x(59 - 4x)))}{112 - x(151 - x(72 - x(14 - x)))}. \tag{3.20}$$

The polynomial $rp(u)$ is first evaluated for $u = 1.60631924$ using single precision IEEE arithmetic, then n results for $rp(x)$ are calculated using MCA for increasing values of x:

$$x = u, (u + \epsilon), (u + 2\epsilon), (u + 3\epsilon), \ldots, (u + n\epsilon), \tag{3.21}$$

with $\epsilon = 2^{-23}$. The difference value $d = rp(x) - rp(u)$ is then calculated for each iteration. Results of both the MCA test and the standard IEEE-754 test can then be compared to determine the difference in result distribution.

3.5.4. *LINPACK*

The LINPACK benchmark determines system performance by measuring the time taken to solve a dense $n \times n$ system of linear equations $Ax = b$ [Ref. 17]. Using this benchmark, performance of the system can be easily measured using an industry benchmark tool, and compared to the performance of an equivalent SW solution. Statistical measurements of the results for x have also been made, and precision testing performed using the benchmark to demonstrate the use of variable precision MCA.s In this work, $n = 300$ was used.

3.6. Results

3.6.1. *System performance and size*

Tables 3.2 and 3.3 provide performance results and logic utilization figures for the MCA co-processor. The primary test used to determine system performance is the LINPACK benchmark, as this is an

Table 3.2.   System performance (measured).

Test Type	MFLOPS	Mean	Std Dev
		Cosine	
SW-FP	990	0.0	0.0
SW-MCA	5.5	0.69	0.000009
HW-MCA	4.7	0.69	0.000009
		Cancellation	
SW-FP	1900	0.0	0.0
SW-MCA	5.2	0.51	0.59
HW-MCA	2.7	0.51	0.59
		Kahan	
SW-FP	1800	0.0	0.0
SW-MCA	4.2	0.000014	0.000033
IIW-MCA	1.5	0.000012	0.000047
		LINPACK	
SW-FP	2350	1.0	0.0
SW-MCA	4.4	1.0	0.0000000004
HW-MCA	3.5	1.0	0.0000000004

industry standard example of a FP benchmark tool. In order to achieve maximum performance the LINPACK benchmark was profiled to determine which functions would provide the most benefit from optimization. These functions were optimized by performing operations of the type $(a * b) + c$ using the co-processor FMA operation. The results of the LINPACK test show that the MCA co-processor achieved a speed of 3.5 MFLOPS for the LINPACK test, and this figure corresponds with the average performance of the system during all test routines. This performance can be compared to the PC performing MCA in software, which can be seen to have achieved an average speed of 4.4 MFLOPS during the LINPACK test, which is again similar to the average speed of 4.75 MFLOPS achieved across all test routines. From this comparison two things are noted, firstly, there is a $200\times$ to $600\times$ decrease in performance for software-based MCA over standard FP. Secondly, the FPGA implementation has comparable performance to the SW implementation. Table 3.3 shows that this performance has been achieved with a $5\times$ increase in logic utilization over a single precision IEEE FP unit capable of performing the same arithmetic operations. Compared with Yeung *et al.* [Ref. 13], this design has similar throughput but a considerable area overhead due to implementing/interfacing each MCA operation separately in order to reduce I/O overhead.

3.6.2. *Improving performance*

The FPGA MCA core implementation has not been fully optimized. Performance is limited by I/O overhead, a conflict between

Table 3.3. System logic utilization.

Operation	DSP48E	FF	LUT	% Inc. (Avg)
ADD	2	595	2067	301%
SUB	2	595	2067	301%
MUL	9	905	2299	725%
DIV	6	4476	7383	761%
FMA	9	1263	3618	383%
TOTAL	28	7834	17434	495%

the -msoft-float and optimization flags in gcc and the maximum clock speed of the implementation. In order to overlap communication with computation the LINPACK benchmark was profiled, with results indicating that 92% of computation time was spent in the **daxpy** subroutines. In addition, analysis of the co-processor I/O overhead showed 80% of execution time spent transferring data and only 20% on computation. The maximum speed-up, S, for LINPACK can be calculated using Amdahl's Law [Ref. 18]:

$$S \leq \frac{P}{1 + f(P - 1)} \tag{3.22}$$

$$\leq \frac{5}{1 + [0.08 * (5 - 1)]} \tag{3.23}$$

$$\leq 3.79, \tag{3.24}$$

where $P = 5$ is the maximum speed increase achievable by minimizing I/O, and $f = (1 - 0.92) = 0.08$ is the ratio of non-daxpy to daxpy computation. This value could be approached since the daxpy operations are vectorizable. Further performance improvements can be obtained using the gcc optimization flags, which were not used in this case due to conflicts with the -msoft-float option. Testing of routines described in this paper by directly modifying the source code showed that a consistent 2× speed improvement is achieved with −O1 optimization. Finally the MicroBlaze maximum clock frequency limits the overall system to a clock frequency to 150 MHz, while the Zynq family of processors recently released by Xilinx are capable of clock frequencies of 800 MHz. In addition Xilinx LogiCORE IP FP operator cores can achieve a maximum frequency of 400 MHz. Thus a conservative estimate for the maximum frequency of an optimized design is 400 MHz, a 2.5 × speed-up over the reported design. The optimizations in this section are orthogonal and, taken together, an additional ∼20 × speed-up may be possible. This would improve the performance of our approach to 60 MFLOPS.

3.6.3. *Error detection*

Figure 3.1 shows the results of two runs of the cancellation test. The histogram on the right shows results using the values given in Sec. 3.5.1 and demonstrates distribution of results for operations susceptible to round-off error. From Table 3.2 it can be seen that the results of this test have a standard deviation of 0.51, this being of the same order as the mean, 0.59. It can thus be concluded that, for the given inputs, the co-processor detected catastrophic cancellation. The distribution of these results can also be compared to the histogram on the right side of Fig. 3.1 which shows results for another execution of the cancellation test using different inputs. From the histogram, it can be seen that the standard deviation and relative standard deviation of the results are much lower and hence, for these inputs, lower sensitivity to rounding error is indicated. The results of the Kahan test are shown in Fig. 3.2. The plot on the right side of the figure shows the results of the Kahan test when performed using standard IEEE-754 FP operations, and as can be seen in the plot, the results do not show random rounding. The plot on the left side of the figure shows results obtained using the MCA co-processor. Comparing these two sets of results it can be

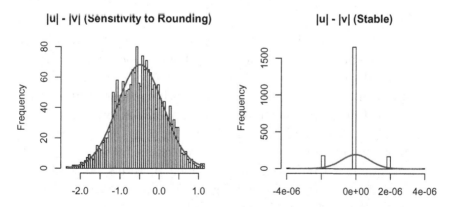

Fig. 3.1. Distribution of results for $[(x + y) + z] - [x + (y + z)]$.

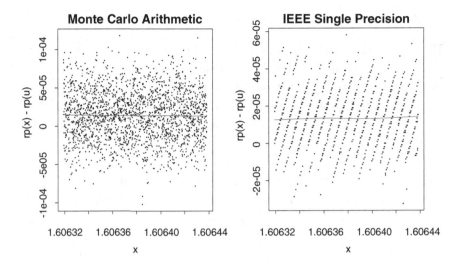

Fig. 3.2.   Kahan test results.

seen that MCA operations performed by the co-processor produce randomly rounded results, and that statistical analysis of these results can be used to determine the sensitivity of the system to rounding error.

3.6.4. *Precision testing*

The final set of testing and results demonstrate the ability of the MCA co-processor to perform variable precision MCA, and to determine the minimum precision required to perform an operation to a specified accuracy. The LINPACK benchmark is executed using $n = 300$ and the virtual precision of MCA operations varied between $1 \leq t \leq 24$. The value of the result vector x is then analyzed to determine the accuracy of the results, which are given in Fig. 3.3. For a required output accuracy (specified in terms of either the minimum number of significant figures or the minimum relative standard deviation), the minimum required machine precision is the corresponding abscissa in Fig. 3.3.

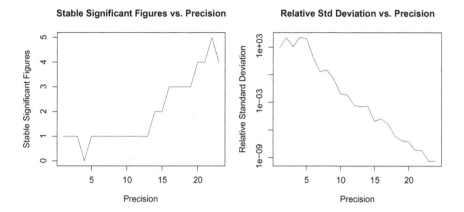

Fig. 3.3. Precision test results.

3.7. Conclusion

A floating-point unit for the run-time detection of round-off errors was designed and implemented using high-level synthesis tools. It was integrated in a MicroBlaze soft processor system, and verified to give results of accuracy equivalent to previously published software implementations. Measurements showed that the performance of the current implementation is similar to an equivalent PC-based SW implementation, and that a further speed-up of 20× is possible.

This work shows that HW accelerated implementations of error detection algorithms can provide accurate measurements of the effects of rounding error while not dramatically impacting performance. Future work will focus on better integration between the FPU and processor, improving performance.

References

1. IEEE. IEEE Standard for Floating-Point Arithmetic, *IEEE Std 754-2008*, pp. 1–70, 2008.
2. D. Goldberg. "Computer Arithmetic", in eds. J.L. Hennessy and D.A. Patterson, *Computer Architecture: A Quantitative Approach*, Morgan Kaufmann, Burlington, MA, USA, 1990.

3. D.S. Parker, B. Pierce, and P.R. Eggert. Monte Carlo Arithmetic: How to Gamble with Floating Point and Win, *Computing in Science and Engineering*, 2(4), 58–68, 2000.

4. A.A. Gaffar *et al.* Unifying Bit-Width Optimisation for Fixed-Point and FloatingPoint Designs, in *Proc. International Symposium on Field-Programmable Custom Computing Machines*, pp. 79–88, 2004.

5. A.A. Gaffar *et al.* Automating Customisation of Floating-Point Designs, in *Proc International Conference on Field-Programmable Logic and Applications*, pp. 523–533, 2002.

6. C.T. Ewe, P.Y.K. Cheung, and G.A. Constantinides. Dual Fixed-Point: An Efficient Alternative to Floating-Point Computation, in *Proc. International Conference on Field-Programmable Logic and Applications*, pp. 200–208, 2004.

7. W. Kahan. How Futile are Mindless Assessments of Round-off in Floating Point Computation. [Online] Available at: http://www.cs.berkeley.edu/~wkahan/Mindless.pdf. [Accessed 23 April 2014].

8. A. Amaricai, M. Vladutiu, and O. Boncalo. Design of Floating Point Units for Interval Arithmetic, in *Proc. Ph.D Research in Microelectronics and Electronics*, pp. 12–15, 2009.

9. M.J. Schulte and E.E. Swartzlander. A Family of Variable-precision Interval Arithmetic Processors, *IEEE Transactions on Computers*, 49(5), 387–397, 2000.

10. J.E. Stine and M.J. Schulte. A Combined Interval and Floating Point Multiplier, in *Proc. Great Lake Symposium on VLSI*, pp. 208–215, 1998.

11. J. Vignes and R. Alt. An Efficient Stochastic Method for Round-Off Error Analysis, in *Proc. Accurate Scientific Computations*, pp. 183–205, 1985.

12. R. Chotin and H. Mehrez. A Floating-Point Unit using Stochastic Arithmetic Compliant with the IEEE-754 Standard, in *Proc. International Conference on Electronics, Circuits and Systems*, vol. 2, pp. 603–606, 2002.

13. J.H.C. Yeung, E.F.Y. Young, and P.H.W. Leong. A Monte Carlo Floating-Point Unit for Self-validating Arithmetic, in *Proc. International Symposium on Field Programmable Gate Arrays*, pp. 199–208, 2011.

14. P. L'Ecuyer. Maximally Equidistributed Combined Tausworthe Generators, *Mathematics of Computation*, 1996.

15. V.G. Oklobdzija. An Algorithmic and Novel Design of a Leading Zero Detector Circuit: Comparison with Logic Synthesis, *IEEE Transactions on Very Large Scale Integration Systems*, 2(1), 1–5, 1994.

16. D.S. Parker. Monte Carlo Arithmetic. [Online] Available at: http://www.cs.ucla.edu/~stott/mca/. [Accessed 23 April 2014].

17. J.J. Dongarra, P. Luszczek, and A. Petitet. The LINPACK Benchmark: Past, Present and Future, *Concurrency and Computation: Practice and Experience*, 15(9), 2003.

18. G.M. Amdahl. Validity of the Single Processor Approach to Achieving Large Scale Computing Capabilities, in *Proc. Spring Joint Computer Conference*, pp. 483–485, 1967.

Chapter 4

The Shroud of Turing

Steve Furber and Andrew Brown

University of Manchester and University of Southampton

We make the case that computing is increasingly compromised by Turing's heritage of strictly sequential algorithms, and the time has come to seek radically different computing paradigms. Whilst we are unable to offer any instant solutions, we point to the existence proof of the vast, distributed, asynchronous networks of computing elements that form brains, and note that FPGAs offer a fabric capable of operating in a similar mode. We offer a computational model — partially ordered event-driven systems — as a framework for thinking about how such systems might operate.

4.1. Introduction

All modern computers are based on an idea first published by Alan Turing as a thought experiment to support his proof of key results about the fundamentals of computability. His concept of the Universal Machine forms the theoretical foundation for the stored-program computer, conceived in practical terms by von Neumann, Eckert and Mauchly, and which first became an engineering reality in the Manchester Baby machine and soon thereafter as a practical computing service supported by Maurice Wilke's EDSAC. At the heart of this idea is the concept of sequential execution: each "instruction" starts with the state of the machine when the preceding instruction has completed, and leaves the machine in a well-defined state for its successor. High-speed implementations bend the actual timing of instruction execution as far as it will go without breaking, but still emulate the sequential model.

The history of advances in computing has revolved around making this very simple execution model go faster, partly through bending the timing of instruction execution as noted above, but mainly through making transistors smaller, faster, and more energy-efficient, all thanks to Moore's Law. This approach has delivered spectacular progress for over 50 years, but then hit a brick wall — the power wall. Since then largely illusory (i.e. marketing) advances in performance have been delivered through multi-core and then many-core parallelism — putting a modest number of sequential execution engines on the same chip, whose potential (i.e. marketing) performance can rarely be realized due to the difficulty inherent in trying to make sequential programs work together in parallel.

The time has come to look again at the fundamentals of computation, and in particular to cast off the shroud of Turing — to abandon sequential instruction execution as the only model of a computation. Alternatives to sequential execution are all around us, including the vast parallel computing resource on a modern FPGA, as studied by Peter Cheung, and the vast complex of biological neurons inside each of our brains. The only minor difficulty is that we do not yet have any general theory of computing on such vast, distributed, networked resources. It's time that changed. Sequential computing is not natural, it's not efficient and, in fact, the only thing it has going for it is that it's easy.

4.2. Sixty Years of Progress

Few would argue that electronics has not burgeoned in the handful of decades since Faraday worried about where and how he might sensibly obtain insulated conductors (silk-covered wire used as stiffening in the manufacture of ladies' hats). This trajectory has been ably documented in many places, and we attempt to do no more than précis the highlights below. However, for a variety of reasons, we argue that this trajectory has begun — or is starting to begin — an inflection. Digital electronics, in all its manifestations, is beginning to converge.

4.2.1. *Von Neumann hardware*

The initial development trajectory of electronic computers is relatively easy to plot: Table 4.1. The arrival of the transistor — invented in 1947, penetrating computer construction in 1955 — changed everything. Power consumption, weight and cost came down by orders of magnitude (but so too did mean-time between failures (MTBF) — the Harwell Cadet Cadet [Ref. 1], introduced in 1955, had no valves, ran at 58 kHz and had an MTBF of 90 minutes). (ENIAC provided remarkable reliability by the simple expedient of never being powered down in the eight years of its life: valves were hot-plugged — in every sense of the phrase — and although programs might be interrupted, the system was never closed down.)

The way forward was clear: faster calculations, more memory, less power, weight and cost. Each of these parameters was attacked vigorously by industry and governments hungry for more performance, and each frontier was successfully and relentlessly pushed back: "By the time that anyone had time to write anything down, it was obsolete." Notwithstanding the diversity of the approaches brought to bear on every aspect of machine development, the legacy of Turing survived almost untouched: machines fetched instructions and data, the former operated on the latter and the results were stored away. The pinch-point of the sequential fetch-execute cycle — if anyone worried about it at all — was way down on the list of rate-limiting attributes that had to be tackled.

4.2.2. *Software*

Arguably the first person to realise the need for some kind of high(er)-level method by which one might program the machines was Zuse [Ref. 2]. He defined the language Plankalkuel ("plan calculus") in 1945, but WWII interrupted his work and it was not implemented until 1998. Plankalkuel contains subroutines, conditionals, loops, floating-point, nested structures, assertions, and exceptions.

Table 4.1. Early development trajectory of the electronic computer. http://en.wikipedia.org/wiki/History_of_Computers.

Name	Date	No. system	Computing mechanism	Programming	Turing complete
Zuse Z3 (Germany)	May 1941	Binary floating-point	Electro-mechanical	Punched 35 mm film stock (no conditional branch)	In theory (1998)
Atanasoff-Berry (US)	1942	Binary	Electronic	Not programmable-single purpose	No
Colossus Mk 1 (UK)	Feb 1944	Binary	Electronic	Patch cables and switches	No
Harvard Mk I — IBM ASCC (US)	May 1944	Decimal	Electro-mechanical	Punched paper tape (no conditional branch)	Debatable
Colossus Mk 2 (UK)	June 1944	Binary	Electronic	Patch cables and switches	In theory (2011)
Zuse Z4 (Germany)	Mar 1945	Binary floating-point	Electro-mechanical	Punched 35 mm film stock	Yes
ENIAC (US)	July 1946	Decimal	Electronic	Patch cables and switches	Yes
Manchester Baby (UK)	June 1948	Binary	Electronic	Stored-program in cathode ray tube	Yes
Modified ENIAC(US)	Sept 1948	Decimal	Electronic	Read-only stored programming	Yes
EDSAC (UK)	May 1949	Binary	Electronic	Stored-program in mercury delay line	Yes
Manchester Mark 1 (UK)	Oct 1949	Binary	Electronic	Stored-program in cathode ray tube and magnetic drum	Yes
CSIRAC (Australia)	Nov 1949	Binary	Electronic	Stored-program in mercury delay line	Yes

Figure 4.1 shows a development chart of the more common high-level languages.

Of these, PL/1 is the first to contain (built in) the notion of a control thread that might be split and recombined under programmer

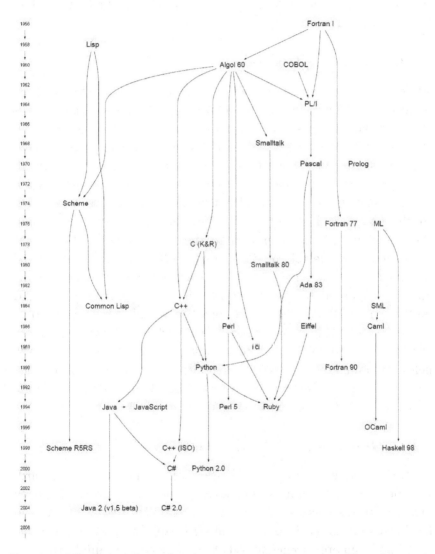

Fig. 4.1.   Inheritance chart of major programming languages. There is a more comprehensive diagram in [Ref. 12] containing over 150 languages, including the CSP-OCCAM and CPL-BCPL-B-C branches.

control. Developed by IBM at their Hursley Research Park, UK, IBM was probably — at the time — one of the few organisations that possessed sufficient computing hardware to sensibly make use of such a construct. Nevertheless, it acknowledged that forcing an instruction stream through a single ALU — however fast — might be something to be avoided if possible.

4.2.3. *Hardware platforms*

After valves came point-contact transistors, then junction devices, then integrated circuits. These offered unprecedented opportunities for those with the resources, but these resources became more and more costly. To overcome this, the concept of an uncommitted wafer was introduced: a silicon die would be created — with all the costs that that incurred — complete up to and including a metallization layer. The metallization layer would not be etched, so the end user had only to supply the final metal mask, and the silicon would be committed to the user's design. The market benefitted from the volume discounts (the NRE would be amortized to something very small), and the user had only to pay for one mask, as opposed to the entire set. (If you wanted to, you could build a computer like this.) More sophisticated platforms emerged — programmable logic arrays, where the underlying silicon could be re-configured multiple times. The final step in this progression is the field-programmable gate array, where pre-designed blocks of complex logic may be reconfigured via a programmable interconnect fabric that is every bit as complicated at the logic blocks it services.

4.2.4. *(Hardware) description languages*

As with computing engines, the growing complexity of the hardware led to an awareness that some form of formal specification mechanism was necessary to get the most out of the hardware, and hardware description languages started to appear. These were *prima facie* similar in appearance to software languages, but unlike these — which allowed the user to dictate the flow of data, and may contain notions of concurrency — hardware description languages contain the explicit notion of time embedded within them, and the user responsibility shifts subtly to require the specification of control, rather than data. Table 4.2 lists a set of common languages. Like software, each has its devotees, but the underlying principle is similar in all of them.

The designer is encouraged to think hierarchically, in terms of functional blocks, and to choreograph the flow of information between them. The notion of a sequential flow of control is either absent, or localized within a specific block at some level in the overall hierarchy.

Meanwhile silicon platforms get bigger, faster, and contain more and more high-level functional blocks. Cores, for example.

4.3. And Then It All Stopped...

The development of many spheres of activity can be characterized by an S-curve, shown in Fig. 4.2.

The curve characterizes many aspects of human activity, from the speed of aeroplanes to the probability of having your car radio stolen. In Phase I, a technology or activity is unexplored, its potential unknown, so few folk enter the field. In Phase II, the potential begins to become realized, and growth becomes rapid as folk realize what can be achieved, and what return on investment can be obtained. Funding, research and exploitation form a positive feedback cycle and the field grows exponentially. Finally, in Phase III, society realises that the boom is over. The outstanding problems become too expensive or just too difficult to solve, and research interest — and with it, the necessary funding — wanes. The product/area has become a commodity. If it breaks, don't fix it, just get a new one. It follows from this that the number of people required to fix it (i.e. the number of people required to understand it) decreases.

Moore's Law [Ref. 3] states that the number of devices on an integrated circuit will approximately double every two years. This is an exponential rate of growth, and exponentials — in any field — are not sustainable in nature. This rate of change predicts that in around

Table 4.2.   Hardware description languages [Ref. 13].

Name	Provenance
Advanced Boolean Expression Language (ABEL)	
Altera Hardware description language (AHDL)	Altera
AHPL	A Hardware Programming Language
Bluespec	Haskell
Bluespec SystemVerilog (BSV)	Bluespec
C-to-Verilog	Converter from C to Verilog
Chisel (Constructing Hardware in a Scala Embedded Language)	Scala (embedded DSL)
CUPL (Compiler for Universal Programmable Logic)	Logical Devices, Inc.
HHDL	Haskell (embedded DSL).
Hardware Join Java (HJJ)	Join Java
HML	SML
Hydra	Haskell
Impulse C	Another C-like HDL
ParC (Parallel C++)	C++ extended with HDL style threading and task communications
JHDL	Java
Lava	Haskell (embedded DSL)
M	Mentor Graphics
MyHDL	Python (embedded DSL)
PALASM	Programmable Array Logic (PAL) devices
ROCCC (Riverside Optimizing Compiler for Configurable Computing)	Free and open-source C to HDL tool
RHDL	Ruby
SystemC	A standardized class of C++ libraries for high-level behavioural and transaction modelling
SystemVerilog	A superset of Verilog, with enhancements to address system-level design and verification
SystemTCL	Tcl
THDL++ (Templated HDL inspired by C++)	VHDL extension with inheritance, templates and classes
Verilog	A widely used and well-supported HDL
VHDL (VHSIC HDL)	

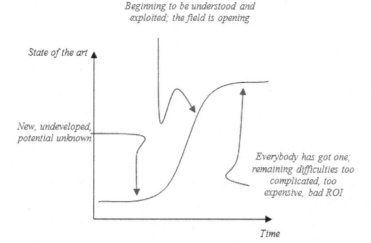

Fig. 4.2.   Universal development "S-curve".

150 years there will be more memory cells on a square centimetre of silicon than there are atoms in the universe.

Something is going to break; what and why?

4.3.1. *The design roadblock*

The resources required to design a state-of-the-art integrated circuit have grown exponentially, following the growth in the transistor resources on the chip itself. Whereas in the 1980s it was possible to design a competitive chip with a small team in a year or so, today's consumer systems-on-chip (SoCs) require teams of hundreds of designers.

Design costs for SoCs are in the $10Ms, so the manufacturing volumes have to be very large to amortize these costs. The number of SoC design starts per year has been going down since 2000; fabless semiconductor start-ups — a popular entrepreneurial model in the 1990s — have all but vanished since 2000 as the investment to break-even has exceeded $100M, which stretches the risk-taking aspect of the venture capital investment too far, leaving SoC design as an enterprise only large companies with established volume markets can undertake. The pricing-out of start-up companies has greatly compromised opportunities for innovation in the microchip business.

4.3.2. *The power roadblock*

Moore's Law is delivered through manufacturing ever-smaller transistors, and as transistors are made smaller they become faster, cheaper and more energy-efficient. But these energy-efficiency gains are more than cancelled out by the increasing number of transistors on a chip, so the chip power budget has grown to the point where it is becoming very difficult, for a high-performance chip, to get the electrical energy in and to get the thermal energy out.

As this trend continues the industry is facing the prospect of "dark silicon" — future chips will simply not be able to operate with all of their functions active at the same time, so at any time several parts of the chip will have to be powered down to avoid melt-down.

4.3.3. *The variability roadblock*

As transistors are made smaller, controlling their operating parameters becomes increasingly problematic. The statistics of variations in the positions of individual atoms within the active region of the transistor are already a significant factor in design. Critical device parameter spreads must be compensated by design techniques, but whereas in the past designers have had to cope with variations of percentages, in the future this will be orders of magnitude larger.

4.3.4. *Device physics*

Atomic-level devices simply don't behave in the manner described by macroscopic models. As transistors shrink towards atomic scales various quantum-level physical phenomena, such as electron tunneling, make themselves felt.

4.3.5. *The interconnect roadblock*

The scaling of transistors to ever-smaller physical sizes has, on the whole, made those transistors better. The same is not true of the interconnect — the wires that join those transistors together. As the transistors have gotten faster, the wires have gotten proportionately slower, and the performance of chips today is limited by the wiring, not the transistors.

On-chip interconnect absorbs both silicon real estate and power budgets; physics places strict limits on causality wave fronts. Off-chip interconnect is a major issue, effectively limiting the ability of external memory to keep many-core processors supplied with code and data.

4.3.6. *The economic roadblock*

The cost of building and running a chip manufacturing facility (a "fab") has grown exponentially alongside Moore's Law, and is now out of reach of many governments and all but the largest multi-national companies.

4.3.7. *May's law*

David May, the British computer architect responsible for the Inmos transputer, made the following observation, now known as May's Law: Software efficiency halves every 18 months, compensating Moore's Law.

The causes responsible for May's Law are a mixture of:

- A shortage of programming skills;
- The tendency of programmers to add too many features;
- Copy–paste programming;
- Massive overuse of windows and mouse-clicks;
- Reliance on Moore's law to solve inefficiency problems.

It is observed, for example, that Microsoft Office 2007 performs (on 2007 hardware) at half the speed of Microsoft Office 2000 (on 2000 hardware) [Ref. 4].

4.3.8. *All in all*

Moore's Law is not a fundamental law; it is not a verifiable hypothesis; it is not an extrapolated prediction (well, it is, actually...); it is not a philosophy. It is a business model — arbitrarily created, trivially discarded. It became a self-fulfilling prophecy, and had significant impact on various roadmaps, but it was based on an unsustainable exponential process. It's time to move on.

4.4. A Condensing of Concepts

With all these obstacles impeding further progress with sequential Turing machines, how do we find a way forward? Let's try to put a few observations and ideas together that can provide a platform for progress in a new direction.

4.4.1. *The world is not synchronous*

Most digital circuits in use today are synchronous: the design is based around the idea that the signal values are binary, and the entire system shares a common notion of time, defined by some clock signal distributed throughout the system. Asynchronous systems, on the other hand, also have binary state values, but there is no notion of an externally dictated passage of time. Sub-circuits use handshaking to choreograph the movement of data throughout the system. Things happen when they happen, and other things predicated on this wait until some "ready" signal is asserted. (One way of looking at this is to assert that every individual part of the circuit has its own, local notion of time, dictated by a local clock, but this clock is driven by the previous process block, and may not necessarily have equally distributed ticks.) Compared to their synchronous counterparts, asynchronous circuits have:

- Lower power consumption, because there are no "wasted" clock transitions;
- A higher data throughput, because block latencies are all local, and there is no requirement to make the circuit tolerant of the worst global latency;
- Lower electromagnetic emissions, because there is no "centre frequency";

- More tolerance of process variation, because timing can be made insensitive to wire and device delays;
- No clock skew problems.

On the other hand, the handshake circuitry itself carries an area and power overhead, and the lack of mainstream acceptance — and hence willingness of large EDA vendors to provide tools — has itself become a vicious spiral. But it is another thread in the growing stream of technologies that are comfortable with things happening in their own time.

4.4.2. *The silicon compiler*

The notion that the behaviour of a system can be captured and translated automatically into a physical structure, in a manner analogous to that in which high-level computing constructs are translated into machine code by a conventional compiler, is called behavioural synthesis, or silicon compilation. Recall that the most significant difference between a software and a hardware language is the addition of temporal information; naively, the difference between a software compiler and a silicon compiler is that the silicon compiler requires knowledge of the temporal cost of each "instruction", so that the dataflow dictated by the user description is not violated by the implementation.

And, of course, the silicon compiler is free to exploit whatever parallelism it can extract from the user description.

4.4.3. *FPGAs — today and tomorrow*

FPGAs [Ref. 5] have been treading their own Moore's Law. The earliest FPGA chips had a few thousand logic blocks, each of which typically consisted of a latch and a few gates of glue logic, that could be configured by an EDA tool only as one complete operation. Today, FPGAs consist of a mass of logic blocks, as before, although the complexity and size of each block has increased, and typically there will many orders of magnitude more of them. The fabric will also be replete with fast IO ports, RAM blocks, DSP blocks, crypto blocks, and explicit cores. The connectivity can be reprogrammed in part or in whole, and sometimes this can be achieved on the fly, under the control of the FPGA itself — truly self-modifying systems.

The key point is that the capability for exploiting parallelism — both fine- and coarse-grained — is multiplying. The user can implement part of the design in a sea-of-blocks, part on (several) explicit fetch-execute cores, part on dedicated high-functionality hardware blocks.

4.4.4. *Multi-core silicon — today and tomorrow*

At the risk of going out of date before the ink has dried, a handful of multi-core technologies illustrate that the multi-core field has its own momentum:

Anton [Ref. 6] is a special-purpose supercomputer consisting of 512 custom ASICs arranged in a high-bandwidth 3-D torus network designed for simulating molecular dynamics (MD) problems. Each ASIC contains a high-throughput compute pipeline tailored for calculating the interactions between pairs of atoms, and a programmable subsystem for calculating FFTs and integrating the particle trajectories. The specific design of the functional units allows Anton to simulate a 62.2 Å cubic box containing 23,558 atoms for 14.5 μs of simulation time in one day of wall-clock time by foregoing general purpose computation support. Network traffic is primarily based around small multicast packets because a high volume of data is constantly streamed through the machine.

Intel have produced a prototype chip that features 48 Pentium-class IA-32 processors, arranged in a 2-D 6×4 grid network optimised for the message passing interface [Ref. 7]. All the processors are coherent and are capable of booting Linux simultaneously but they are designed as coprocessors rather than a stand-alone system. A special message-passing buffer shared by processors and the introduction of a message-passing memory type implements communication between the cores using caches.

Centip3De is a 130 nm stacked 3-D near-threshold computing (NTC) chip design that distributes 64 ARM Cortex-M3 processors over four cache/core layers connected by face-to-face interface ports [Ref. 8]. The 3-D network provides a large area reduction inside the chip but cannot be used for chip-to-chip networks. Four processors share instruction and data caches that are part of a 3-D network that permeates the device to ensure coherency and to provide access to the 256 MB of system memory.

The Swizzle-switch network [Ref. 9] is a 128 bit 64-input 64-output single-stage swizzle-switch network (SSN) which is similar to a crossbar switch but also supports multicast messages. A least-recently granted (LRG) arbiter is embedded in the network fabric to implement channel switching that must be explicitly released by the originating master. It is capable of providing on-chip bandwidths as high as 4.5 Tbps but the large wire count makes it impractical for chip-to-chip networks.

TILE64^{TM} is a chip-multiprocessor architecture design that arranges 64×32 bit VLIW processors in a 2-D 8×8 mesh network that supports multiple static and dynamic routing functions [Ref. 10]. High-bandwidth input/output (I/O) ports allow the TILE64 to act as either a coprocessor or a system processor optimised for streaming data tasks such as data-center network routing or deep packet inspection algorithms. A 5-port crossbar-switch-based wormhole router attached to each processor tile allows caches to be connected together and to the system memory to provide a mechanism for coherence.

As the size of parallel systems increases, the proportion of resource consumption (including design effort) absorbed by "non-computing" tasks (communications and housekeeping) increases disproportion-ally. Architectures that sidestep these difficulties with unconventional approaches are gaining traction in specialised areas.

4.4.5. Where's it all heading, then? Convergence...

Our hypothesis, then, is that future progress in computing is dependent on models of computation that abandon Turing's heritage of sequential algorithmic execution, and instead see computation as a property of a vast distributed, asynchronous, non-deterministic network of agents. On this basis it will be hard — in some number of years' time — to tell the difference between a state-of-the-art FPGA and a state-of-the-art multi-core computer platform.

When we get to the point that — for all practical purposes — the hardware is free, we may as well make the individual point cells as complicated as we like. It may seem anathematic to use an ARM core as a dedicated arithmetic operator, but if we have an infinite number of cores available and the operation is not on the critical path, why not?

As a way to explore this space, let's build a machine with a million cores on it, and see what it can do.

4.5. We're Closer than You Might Think: Computing on Vast, Distributed Networked Resources

There are already many examples of systems that implement some form of information processing on vast, distributed networks. These can be found in nature and in engineering. Natural systems include:

- The brain, of course;
- Colonies of social insects;
- Businesses and industries;
- Human society?

Engineered examples include:

- The internet.

4.5.1. *How does Nature do it? Of brains, neurons and synapses*

The brain is the prime example in nature of the sort of system that supports our hypothesis that meaningful (albeit non-deterministic) computation can be performed on vast distributed networks of asynchronous agents. The fundamental drawback here is that science cannot yet offer anything approaching a complete description of how this organ, so central to the lives of each of us, carries out its vital functions.

Brains are composed of neurons, and neurons are, like logic gates, multiple-input single-output devices. True, they tend to have a lot more inputs: logic gates typically have two, three or four inputs, whereas neurons typically have 1,000, 10,000 or even sometimes 100,000 inputs, so there is a difference of scale. Neurons connect through synapses, and communicate principally by sending impulses when their inputs resemble a pattern they are tuned to. Synapses are plastic — they change under the influence of the activities of the pre- and post-synaptic neurons — and the wiring diagram of the neural circuit is itself subject to change over time (unlike most electronic circuits, although dynamically reconfigurable FPGAs can change their effective wiring at run-time).

Brains also display structural regularities — neuroscientists talk in terms of the 6-layer micro-architecture of the cortex, which looks similar throughout, including at the rear of the brain where it supports low-level image processing that is to some degree understood, and at the front of the brain where it supports natural language processing and higher levels of thought, where we currently have little clue as to what is going on. Yet the fact that these very different processes run on the same substrate must be telling us something about the algorithms that are in use? Likewise the 2-D nature of the cortex and the general use of 2-D topographic maps suggests that the 2-D map may be a general principle of operation, though this is little more than a hypothesis at this stage.

4.6. A Computer Engineer's Approach

The world of computing is obliged to move into unknown territory. The only existence proof that there is something out there that works is the brain, but there is the slight difficulty that we have no idea how the brain works. The computer engineer's approach is then to ask if we can use what we do know about building computers to help understand the inner workings of the brain. This line of thought has led to the SpiNNaker project, an attempt to build a massively parallel computer inspired by, and optimized to model, what is known about the detailed working of the brain.

4.6.1. *SpiNNaker*

SpiNNaker [Ref. 11] (Fig. 4.3) is a multi-core message-passing computing engine based upon a completely different design philosophy from conventional machine ensembles. It possesses an architecture that is completely scalable to a limit of over a million cores, and the fundamental design principles disregard three of the central axioms of conventional machine design: the core-core message passing is non-deterministic (and may, under certain conditions, even be non-transitive); there is no attempt to maintain state (memory)

Fig. 4.3.   A 48-node (864-processor) SpiNNaker board.

coherency across the system; and there is no attempt to synchronize timing over the system.

Notwithstanding this departure from conventional wisdom, the capabilities of the machine make it highly suitable for a wide range of applications, although it is not in any sense a general-purpose system: there exists a large body of computational problems for which it is spectacularly ill-suited. Those problems for which it is well-suited are those that can be cast into the form of a graph of communicating entities. The flagship application for SpiNNaker — neural simulation — has guided most of the hard architectural design decisions, but other types of application — for example mesh-based finite-difference problems — are equally suited to the specialized architecture.

There is a low-level software infrastructure, necessary to underpin the operation of the machine. It is tempting to call this an operating system, but we have resisted this label because the term induces preconceptions, and the architecture and mode of operation of the machine does not provide or utilize resources conventionally supported by an operating system. Each of the million (ARM9) cores has — by necessity — only a small quotient of physical resource (less than 100 kbytes of local memory and no floating-point hardware). The inter-core messages are small ($<=72$ bits) and the message passing itself is entirely hardware brokered, although the distributed routing system is controlled by specialized memory tables that are configured with software. The boundary between soft-, firm- and hardware is even more blurred than usual.

SpiNNaker is designed to be an event-driven system. A packet arrives at a core (delivered by the routing infrastructure, and causes an interrupt, which causes the (fixed size) packet to be queued. Every core polls its incoming packet queue, passing the packet to the correct packet handling code. These packet event handlers are (required to be) small and fast. The design intention is that these queues spend most of their time empty or, at their busiest, containing only a few entries. The cores react quickly (and simply) to each incident packet; queue sizes $\gg 1$ are regarded as anomalous (albeit sometimes necessary). If handler ensembles are assembled that violate this assumption, the system performance rapidly (and uncompetitively) degrades.

4.7. Partially Ordered Event-Driven Systems (POEDS)

Here we attempt to begin to develop a formal system model that can be used to describe the operation of biological neural systems (such as the brain) and computational models of such systems. This work is motivated by a desire to find useful ways to think about information processing in the brain, and by a desire to produce a formal semantics that can underpin reliable operation of the SpiNNaker machines.

We introduce three models at different levels of abstraction, progressing from the biology of neural systems down to the details of the SpiNNaker machine.

4.7.1. *A hybrid-system model*

The system is a set of dynamical processes that communicate purely though event communications using a set of event channels. Each dynamical process evolves in time under the influence of received events; typically a process will depend on a subset of the event channels, not all of them.

Each event channel carries events that are either generated by a process, or come from the environment, and go to all the other processes that depend on them. Normally an "event" is a pure asynchronous event that carries no information other than that it has occurred, so an event channel can be thought of as a time series of identical impulses.

We can consider the event channel to be instantaneous, so that events arrive at all of their destinations at the same time that they are generated by their source process, though causality allows us to view this as "after" they are generated, albeit by a vanishingly small delay. Likewise, if an incoming event causes a process to generate an outgoing event this causality is captured by the output being "after" the input.

The outputs from the model are simply a subset of the total set of events.

4.7.2. *Biological neurons*

Biological neurons are complex living cells that have a cell body (the soma), a single output (the axon) that carries action potentials, and a complex multi-branched input structure (dendrites) that collect inputs. The axon from one neuron couples to the dendrite of another through a synapse, which is a complex adaptive component in its own right.

Action potentials are sustained and propagated by electro-chemical processes in the axon that allow them to be viewed as pure asynchronous events.

Long axons incur significant delays, but these can be rolled into the transmitting and/or receiving process. Where there are different delays from a single source to different targets, for example a short delay to proximal targets and a long delay to distal targets, the hybrid-system model allows this to be captured either by different delays in the receiving process or by the source transmitting separate events with different source delays, or some combination of these.

We therefore claim that the hybrid-system model captures the essential features of biological neurons that exchange information principally through action potentials.

Action potentials are not the whole story, however. Some neurons produce chemical messages, for example dopamine, that modulate the activity of other neurons within a physical region. Some neurons make analogue dendritic connections with their neighbours. These phenomena are outside the hybrid-system model, but we hope that their principal effects can be captured through back-channel processes of some sort.

In addition, biological systems do not have static connectivity — they develop and grow, gaining and losing neurons and connections to their "event channels". But these happen slowly relative to the real-time information flow, and again we hope to implement connectivity changes through back-channel processes as and when we get to that point.

Biological systems are also very noisy, but we can accommodate this by using noisy processes.

4.7.3. *An abstract computational model*

We cannot compute a continuous process exactly as in the hybrid-system model, so for efficiency it is important to approximate the process in some way. Most neuron models are some form of system of differential equations, so it is common practice to compute these using a form of Euler integration over discrete time-steps.

For real-time modeling, the Euler integration can be implemented by introducing an additional "time-step" event. Now time is just another, regular (e.g. 1 ms) event, from an external source, and time can be removed from the model.

We can, at least in principle if our computer is sufficiently fast, ignore the time taken for a process to handle an event. Each event is handled as it arrives, and each process is simply a set of rules defining how that process's state is changed by every possible input event. This is the event-driven aspect of POEDS.

It is clear that a process is active only in response to an input event, and therefore any output events it generates must also occur at the same time as (though causally after) an input event. Note that this does not preclude an internal time delay between a neural input and the output it causes: the input can change the state of the process, which then progresses through several time-step events before producing an output. But the output will eventually be produced in response to, and at the same time as, a time-step event. As the only representation of time in the system is the time-step event, time is discretized.

Since the time-step event connects to many (if not all) processes, there may be many events generated just after it. These events, from different processes, have no implicit order. This gives rise to the partially ordered aspect of POEDS. Each process to which some of these concurrent events are inputs will impose an arbitrary order on their reception (at notionally the same time), and as a consequence the system behaviour is non-deterministic at this point.

4.7.4. *A SpiNNaker computational model*

SpiNNaker is a massively parallel system with an interconnect fabric designed specifically to convey events generated by a program running on one processor to all of the processors to which that event is an input. The SpiNNaker fabric must initially be configured to put the necessary connections in place, but once so configured the hardware looks after the event connections. Processors then receive events intended for them and issue events with no knowledge of where they are destined to go.

Unfortunately the processors on SpiNNaker aren't infinitely fast, so a process takes a finite time to complete its response to an input event. While it is running another event may arrive, demanding pre-emption. The time-step event may not be synchronized across the machine (although near synchronization might be possible using a technique such as fire-fly synchronization).

A further complication is that SpiNNaker processors keep some of their state in off-chip SDRAM, access to which incurs high latency costs. In general we aim to hide this latency by exploiting DMA subsystems attached to each processor to handle SDRAM transfers while the processor gets on with other stuff.

These (and other) niceties apart, SpiNNaker aims to implement the abstract computational model as faithfully as it can, subject to all of the constraints of the physical system, delivering a reasonably efficient solution, and minimizing energy consumption.

SpiNNaker models may attempt to implement the abstract computational model faithfully, in which case they will aim to synchronize the (notional) 1 ms time-step across the machine and complete all the work in every 1 ms to stay in lock-step across the machine. In this case the peak process load must complete within the 1 ms for correct operation. Or they may adopt an asynchronous model where there is no attempt to align a 1 ms period in one process with that in another, in which case the average process load must complete within 1 ms for correct operation.

4.7.5. *Spiking neurons on SpiNNaker*

Each processor on a SpiNNaker machine handles one process, where each process models a number of neurons. As incoming events from other processes are very similar they are handled by one event handler. The simplest model of a SpiNNaker process then handles two event types:

1. Incoming neuron event: locate & process synaptic data, updating local neural state accordingly.
2. Time-step event: perform Euler integration step for all local neurons, possibly generating outgoing events.

As an implementation detail the neuron event handler will usually invoke a DMA transfer to bring the synaptic connectivity data in from SDRAM, but as this is internal to the process we hope to hide the DMA as much as possible from the application code.

This model does not handle the important aspect of synaptic plasticity, but already creates some interesting data consistency issues if event 2 occurs while the (fairly long) event 1 process is running and pre-empts it. These data consistency issues are avoided if no input is allowed to affect state that is used in the current time step, which amounts to imposing a minimum axonal delay of 1 ms.

In general a SpiNNaker implementation will use a very simple real-time kernel of some sort, with drivers for the event communication system, DMA, etc. It will need queue management, priority scheduling, buffer overflow procedures, and so on. But it will maintain a strongly event-driven nature, spending any idle time in a low-power wait-for-interrupt state.

4.8. Conclusions

We have come a long way building computers on Turing's sequential foundations, but we've reached the end of the road. It's time to try something new, but what? The brain is an existence proof of a different way to process information, but we don't know how it works. Computers can help in the scientific quest to understand the brain, and SpiNNaker has been designed with that end in mind. It offers a different way to think about computation. Where that will lead, we do not know.

References

1. S. Lavington. *Early British Computers*, Manchester University Press, Manchester, UK, 1980.
2. K. Zuse. Über den allgemeinen Plankalkül als Mittel zur Formulierung schematischkombinativer Aufgaben, *Archiv der Mathematik*, 1(6), 441–449, 1948.
3. G. Moore. Cramming More Components onto Integrated Circuits, *Electronics*, 38(8), 114–117, 1965.
4. R.C. Kennedy. Fat, Fatter, Fattest: Microsoft's Kings of Bloat. [Online] Available at: http://www.infoworld.com/t/applications/fat-fatter-fattest-microsofts-kings-bloat-278?page=0,4,. [Accessed 23 April 2014].
5. P.Y.K. Cheung, G.A. Constantinides, and J.T. de Sousa. Guest Editors Introduction: Field Programmable Logic and Applications, *IEEE Trans Computers*, 53(11), 1361–1362, 2004.
6. D.E. Shaw *et al.* Anton, a Special-purpose Machine for Molecular Dynamics Simulation, *ACM SIGARCH Computer Architecture News*, 51(7), 91–97, 2008.
7. J. Howard *et al.* A 48-core IA-32 Message-passing Processor with DVFS in 45 nm CMOS, in *Proc. International Solid-State Circuits Conference Digest of Technical Papers*, pp. 108 100, 2010.
8. D. Fick *et al.* Centip3De: A 3930DMIPS/W Configurable Near-threshold 3D Stacked System with 64 ARM Cortex-M3 Cores, in *Proc. International Solid-State Circuits Conference*, pp. 190–192, 2012.
9. S. Satpathy *et al.* A 4.5 Tb/s 3.4 Tb/s/W 64×64 Switch Fabric with Self-updating Least-recently-granted Priority and Quality-of-service Arbitration in 45 nm CMOS, in *Proc. International Solid-State Circuits Conference Digest of Technical Papers*, pp. 478–480, 2012.
10. S. Bell *et al.* TILE64 TM Processor: A 64-Core SoC with Mesh Interconnect, in *Proc. International Solid-State Circuits Conference Digest of Technical Papers,* pp. 88–89, 2008.
11. S.B. Furber *et al.* Overview of the SpiNNaker System Architecture, *IEEE Transactions on Computers*, 62(12), 2454–2467, 2012.
12. Diagram & History of Programming Languages. [Online] Available at: http://rigaux.org/language-study/diagram.html. [Accessed 23 April 2014].
13. Hardware Description Language. [Online] (Updated 3 April 2014) Available at: http://en.wikipedia.org/wiki/Hardware_description_languages. [Accessed 23 April 2014].

Chapter 5

Smart Module Redundancy — Approaching Cost Efficient Radiation Tolerance

Jano Gebelein, Sebastian Manz,
Heiko Engel, Norbert Abel, and Udo Kebschull

*Infrastructure and Computer Systems for Data Processing (IRI),
Goethe-University Frankfurt*

This chapter deals with the problems when operating FPGAs in radiation environments and presents practical realizations that implement techniques to mitigate radiation effects.

Hereby the focus is on smart module redundancy which replaces conventional TMR (triple module redundancy) and aims for higher cost efficiency. The basic idea is to take a closer look at the FPGA design and classify its parts. Each class is then protected in its own most efficient way.

Current implementations are used in read-out boards by the *Compressed Baryonic Matter* collaboration in test setups for high-energy physics detectors at GSI/FAIR in Darmstadt, Germany. These physics experiments are challenging in two ways: they come with very high radiation rates and demand very high cost efficiency.

In summer 2009 the last author of this chapter had the opportunity to join Peter Cheung's research group for a short research visit at Imperial College. During his stay in London, he got in touch with PhD students, post docs and lecturers who worked on topics like chip aging effects, fault tolerance, and characterizing field programmable gate arrays (FPGA) for better optimization of timing and resource usage. Many of the ideas discussed with Peter and his students had an influence on the radiation mitigation techniques described in this chapter.

The authors' strong focus on radiation tolerance is motivated by their involvement in high energy heavy ion collider experiments like the ones currently run at CERN in Geneva/Switzerland and at GSI/FAIR in Darmstadt/Germany. Here, FPGAs are used for the read-out of experiment data, first stages of data reduction, and feature extraction of sensor data. Experiment data needs to be reduced in the very early stages due to extremely high data rates, and therefore it is necessary to operate the FPGAs as close as possible to the collision point. While space applications in low-earth orbit (LEO) usually need to handle a few errors per month, detector applications in heavy ion experiments need to handle up to ten errors per second.

This chapter focuses on ways to protect an FPGA design against radiation-based single event upsets (SEU) in a cost-efficient way. The gained insights show that the implemented techniques help to securely operate standard FPGAs in high-radiation environments.

5.1. Introduction

FPGAs are well-known devices in the world of reconfigurable hardware. They are physically built of various logical components like look-up tables (LUT), multiplexers, and flip-flops, supplemented by a couple of manufacturer specific parts like clock managers or highly specialized digital signal processors. A chip-wide interconnection routing network is provided for internal crosslinking and signal transfer between all of these components. Nowadays, two major commercial FPGA types are available on the market: Flash-based non-volatile and static random-access memory (SRAM) based volatile CMOS architecture chips. Both models provide many well-known advantages but nevertheless some major disadvantages, especially when used in radiation-susceptible environments such as avionics, space applications or particle accelerators. Therefore, FPGA manufacturers like Xilinx provide chip features which allow partial reconfiguration that can be used to refresh configuration data like routing information and static LUT content at runtime: so-called scrubbing [Ref. 1].

The FPGA material's interaction with ionizing particles is founded in the SRAM architecture's general characteristics. The penetrating radiation causes the semiconductor chip's doped silicon to change its electrical properties. This physical separation of electron-hole pairs results in spontaneous single event effects (SEE). These SEEs can for example change a signal temporarily from logic 0 to logic 1, or vice versa. If this signal hazard is picked up by a flip-flop, the temporary hazard gets stored and thus becomes a permanent SEU [Ref. 2]. In general, these errors may lead to spontaneous unexpected system behavior, e.g. finite state machines (FSM) entering undefined states. In the worst case they can lead to a total system halt, known as single event functional interrupt (SEFI).

To mitigate these radiation effects, various approaches are available: extensive hardware shielding as well as the use of radiation hardened materials for manufacturing up to a triple-chip design complemented by an external voter circuit. None of these approaches is applicable for use in particle accelerators, because of the tremendous size or financial requirements for higher quantities. Another option is the integration of fault tolerance in a single FPGA using spatial as well as temporal circuit redundancy. However, this can be extremely resource-intensive as discussed in the following section.

5.2. State of the Art

According to the well-known problem of SRAM CMOS architecture's susceptibility to radiation, some major circuit design hardening technologies were developed [Refs. 3–5]. One of their major disadvantages is the inapplicability to circuits providing re-programmability features like FPGAs with universally and evenly spread configurable logic blocks. Also the associated increased cost factor is not negligible. Thus, very specialized shielded radiation-hard materials have been and currently are developed for military and space grade FPGAs, realized within the Xilinx Virtex QPro II, 4 and 5 series, complemented by the static scrubbing feature.

As these chips are not available for custom non-US applications, commercial off-the-shelf (COTS) devices are provided with special total ionizing dose (TID) annealing techniques and SEE failsafe combinations of logic blocks. Basically, such ambitions in securing the logical hardware design layer can be realized by three different approaches: spatial redundancy, temporal redundancy and various combinations of both. Spatial redundancy features synchronous data sampling of combinatorial logic at multiple routes to mitigate SEUs. The addition of adjacent voters is required to analyze processed data values. Well-known candidates using this principle are dual and triple modular redundancy (DMR/TMR), mostly accompanied by error detection and correction codes (EDAC). Temporal redundancy enables a single combinatorial logic circuit to be sampled at multiple times. Additional voter circuitry compares all of the results and decides whether an error occurred or not.

The combination of both spatial and temporal redundancy leads to SEU immunity at the price of multiplied resource requirements and timing decrease. Complementary fault tolerance techniques like self-replicating temporal sampling [Ref. 6] try to re-use chip resources to assure data integrity at lower costs. Furthermore, multiple bit upsets can be handled using weighted voting techniques [Ref. 7]. However, some of these additional design methods have been proven to increase susceptibility to radiation instead of reducing it when used in a generalized context [Ref. 8].

These days, TMR in combination with scrubbing has become an unofficial industry standard for adding fault tolerance to logic circuits. FPGA vendors like Xilinx provide tools that take a given FPGA design, triple it and add voters as part of the place and route process. The advantage of this method is that it is fully automated. The huge drawback of this approach is the area consumption that comes with it. Spatial requirements typically grow up to six times the original size. This is a big problem regarding cost efficiency since using a six-times bigger FPGA can easily translate to more than 20 times the cost, depending on the current market situation.

5.3. Smart Module Redundancy

Our basic approach is to take a closer look at a given design and its components and to ask for each component individually: what do we actually need to do to make it radiation tolerant?

It turned out that there are principally three classes of components. Each class demands its own mitigation technology:

(1) Data Path — buses, registers and FIFOs that transport, store and process a data stream.
(2) I/O — specific FPGA components used to communicate with the outside world.
(3) Control Logic — registers, FSMs or even embedded processors used to monitor and control the data stream processing.

5.3.1. *Data path*

Most data paths are not limited to a single FPGA, but spread across several devices. Using a communication medium like optical fiber, these devices can even be located far away from each other. This naturally comes with the danger of media-introduced errors. Thus, nearly all communication protocols (such as Ethernet) contain redundancy mechanisms like a parity bit or a CRC that allow the detection of introduced errors. Normally, these protection mechanisms focus more on the world outside the FPGA. A typical FPGA-based Ethernet switch would for example check the CRC of an incoming stream, strip the CRC, and re-attach a CRC on the outgoing stream.

In order to make the data path radiation-hard, it is sufficient to keep these protection mechanisms and use them inside the FPGA (for example enhance the internal data stream by a CRC). This makes it possible to detect and to correct SEUs without the need for TMR — which finally leads to a radiation-tolerant system that is much smaller than a fully triplicated system. This is especially important, since the data path is generally the biggest part of the FPGA.

For example, a design implementing a 128-bit data stream with 20 pipeline stages and 3 FIFO stages consumes 320 registers and 3 FIFOs. The TMR version of this data stream including voters uses about 1,250 registers and at least 9 FIFOs. Instead, adding a 32-bit CRC to the 128-bit data stream leads to a consumption of 400 registers. Today's FIFOs provide additional bits for CRCs and other protection mechanisms, so it is quite safe to assume that the CRC design still only needs three FIFOs. In this simple case example, smart module redundancy saved 850 registers and 6 FIFOs.

5.3.2. I/O

The FPGA's I/O is in many ways highly comparable to the data path. The basic rule for I/O is: if one places an FPGA in a radiation environment and lets it communicate with other devices, mitigation technologies always have to focus on the system as a whole, not only on the exposed FPGA. In the experiments at GSI/FAIR the radiation-exposed FPGA communicates via optical fibre with an FPGA located in a shielded control room. The communication protocol between these two FPGAs has been designed in a way that handles transmission errors. Therefore it is not really important if these transmission errors are caused by an SEU in the FPGA's output buffer or by poor signal quality. Most conventional communication protocols (such as TCP/IP) are tailored to handle these transmission errors. Thus, a very cost-intensive triplication of the I/O pins is not necessary. At GSI/FAIR, the communication protocols were additionally specified to be robust against a temporary complete device failure happening on the exposed FPGA.

5.3.3. Control logic

The most complex part of the FPGA design, in terms of mitigation methodologies, is control logic. Here, SEUs can cause effects that are both hard to predict and hard to monitor. The basic reason behind this is: every monitoring or mitigation logic implemented on the FPGA introduces a new radiation-sensitive element that can itself become a victim to SEUs — and thus may cause more problems than it actually solves. Examples for effects caused by SEUs in control logic are:

- Sequence counters that jump to a non-sense value.
- Status bits that change content (e.g. fifo_empty).
- FSMs that enter an undefined state or violate the defined state transition matrix.

Trying to find a clever way to protect control logic has proven to be very complex — and extremely prone to misjudgement, ending in bad surprises [Ref. 8]. Thus, our observation is that the safest way to protect the control logic is to make use of automated DMR or TMR. The cost of this is not too high, since control logic makes up only a small part of a typical FPGA design. For example, in the case of GSI/FAIR's GET4 Read-Out Controller [Ref. 9] it only represents 10% of the entire design.

Having said that, long-term experiments at GSI/FAIR revealed that one cannot fully guarantee that the protection mechanisms will work. If an SEU hits a clock net or another critical part of the FPGA, even the best mitigation technology can fail. Due to this, it is very important to have radiation-hardened watchdogs in place (normally implemented on a radiation-hard device outside the FPGA) that are permanently monitoring the FPGA and are able to reset the whole SRAM chip if necessary. The communication protocols have to be robust against this temporary device failure (as described above).

The most complex control logic scenario is the utilization of a CPU. In many cases this simply leads to not using a CPU in radiation environments in the first place — but this is not always possible. CPUs come with the advantage of high flexibility and very high adaptability. Moreover, software running on a CPU can execute very complex tasks with a much lower resource consumption than a comparable pure-firmware solution. Hence, talking about cost efficiency means talking about CPUs. Due to this, our workgroup developed a fault-tolerant VHDL soft-core CPU fully compatible with the MIPS R-3000 architecture. It is described in detail in Sec. 5.4.2.

5.4. Test Setups and Measurements

This section presents two test scenarios in which COTS FPGAs have been exposed to high radiation doses in order to prove the effectiveness of smart module redundancy. The first setup focuses on design classification and the corresponding mitigation strategies. The second setup focuses on a radiation-tolerant CPU.

5.4.1. *Test setup I: design classification [Ref. 10]*

The main objective of this test was to measure the radiation mitigation capability of smart module redundancy on an *operational* detector read-out design running on an SRAM-based FPGA in a high radiation level environment.

The tested design is the firmware for the FPGA-based read-out controller for the GET4 ASIC [Ref. 9]. This ASIC is the planned time to digital converter chip for the time of flight (ToF) detector in the compressed baryonic matter (CBM) experiment that is currently constructed in Darmstadt, Germany as part of the Facility for Antiproton and Ion Research (FAIR).

The experiment was carried out at the Cooler Synchrotron (COSY) particle accelerator at the Forschungszentrum Jülich in Germany. The accelerator was configured to provide a 2.1 GeV/c proton beam with a particle rate in the order of $10^7 \, \text{s}^{-1} \cdot \text{cm}^{-2}$.

The beam particle rate was roughly 1,000 times higher than the expected fast hadron flux in the environment of the detector. The expected failure rate is significantly lower for a single FPGA, however, since the number of FPGAs that are required to read out the full CBM-ToF detector is also on the order of 1,000, the failure rates measured in this beam test are very well comparable to the failure rate that is expected for the fully equipped CBM-ToF detector.

5.4.1.1. *The device under test*

For the beam test we used the CBM read-out controller board which is widely used by the CBM collaboration [Ref. 11]. The board is equipped with an SRAM-based Xilinx Virtex-4 FX20 FPGA as core data processing device. A small, flash-based Actel ProASIC3 A3P125 FPGA in conjunction with an on-board flash memory is used as configuration controller for the Virtex-4. The Actel is programmed to continuously scrub the Virtex-4 FPGA configuration (blind scrubbing).

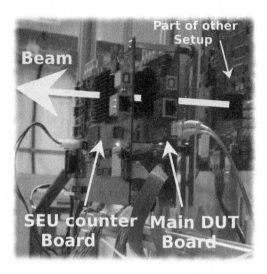

Fig. 5.1. Depiction of the setup. Two boards are mounted in the beam line, one of them represents the main device under test (DUT) which is running the read-out firmware, the other one is a reference board with an identical FPGA counting the SEUs.

5.4.1.2. *The firmware under test*

The firmware under test was not an academic test design, but the actual core of an operational design which is being used to read out real detector front-end electronics (the GET4 chip). The logic was initially classified and then modified to operate in radiation environments:

- Control Logic — critical parts of the control logic have been triplicated and state machines were designed to recover to normal operation even if they entered an undefined state for any reason [Ref. 12].
- I/O — the communication protocols were specified to be robust against temporary device failure.
- Data path — a CRC checksum has been added to the data path.

5.4.1.3. *Test procedure algorithm*

The test procedure which was running on the data acquisition (DAQ) computer has been specifically designed to evaluate the efficiency of the scrubbing technique. Figure 5.2 illustrates the sequence of the algorithm.

First, in the `Init` step, some essential parameters were logged. This included the SEU rates of both devices in the beam which were recorded in parallel for three minutes in order to cross-check whether the SEU rates of both devices were indeed the same. Next, the processing of the main loop began with `Take Data`, dumping three seconds ($\sim$15 MB) of raw data to the hard disk for later offline analysis. Following this, in the `Readback` step, the configuration memory of the SEU counter device was read out to record the current SEU rate.

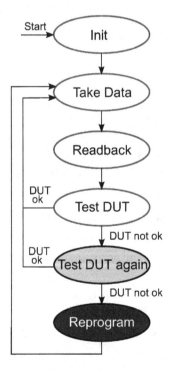

Fig. 5.2. The key steps of the test procedure (`Test DUT`) are performed twice to allow the DUT to be repaired by scrubbing.

Finally in **Test DUT** the functional status of the device under test (DUT) was inspected based on the analysis of 2 000 data samples. Since the data samples were sent by a deterministic data generator, their consistency could easily be evaluated. If all data samples were flawless, the DUT was considered to be fully operational and the test procedure would continue with the next loop iteration again with **Take Data**. If one or more corrupt data samples were detected the DUT was considered to be *not* fully operational, most likely due to a critical SEU. In this case, instead of instantly reprogramming the device, the very same data consistency check was run again (**Test DUT Again**) to allow the setup to be repaired by the scrubbing technique. A scrubbing cycle is much shorter ($\sim 80\,$ms) than the time between the two consistency checks ($\sim 1\,$s). Only if the second check had also failed the algorithm entered the **Reprogram** state where the whole setup would be completely reset. This method made sure that the system could recover even from a SEFI.

5.4.1.4. *Results*

Figure 5.3 shows a direct impression of the values recorded during the beam time. Both diagrams show the results of a three hours run, where scrubbing was disabled (Fig. 5.3(a)) or enabled (Fig. 5.3(b)). Please keep in mind that in order to accelerate the tests, the FPGA has been placed in the middle of the particle beam and thus the beam particle rate was roughly 1,000 times higher than it will be for the actual setup at GSI/FAIR.

The plots marked with squares in Fig. 5.3 represent the time period since the last persistent error has been detected and the test procedure has entered the **Reprogram** state. Significant differences exist between the run with enabled and disabled scrubbing. Without scrubbing the system fails within about one minute and is only stable during the time of technical stops. However, if scrubbing is enabled, the system can survive for several minutes. This shows that scrubbing in combination with smart module redundancy has reduced the downtime of the setup by a factor of almost 50.

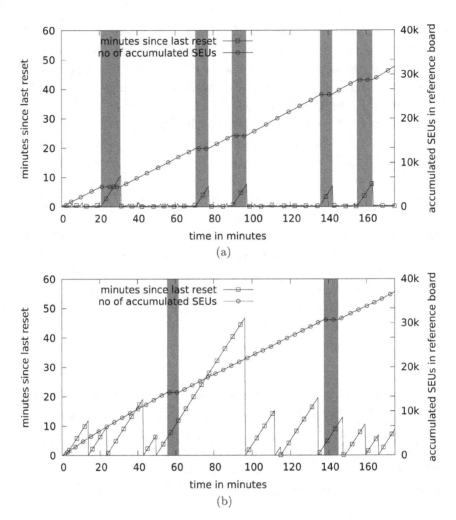

Fig. 5.3.   The plot marked with circles refers to the number of SEUs collected in the reference board. The plot marked with squares shows the time period since the last full reset of the setup (Reprogram state in Fig. 5.2). During the highlighted time slots the beam was shut down for technical reasons. (a) Scrubbing is disabled. Full reset of the setup required in less than a minute. The setup is only stable when beam turned off. (b) Scrubbing is enabled. The setup runs stably for several minutes.

The analysis of the data obtained in **Take Data** step shows that scrubbing in combination with smart module redundancy also significantly improved the data quality. When scrubbing was not applied, about 7% of the data became corrupted, whereas when scrubbing was enabled, only 0.03% of the data was inconsistent.

5.4.2.   *Test setup II: a fault-tolerant CPU [Ref. 13]*

In order to be able to use a CPU in radiation environments, a fault-tolerant VHDL soft-core CPU, fully compatible with the MIPS R-3000 architecture and instruction set [Ref. 14] has been developed.[a] This enables a standard GNU GCC MIPS cross compiler to be used for software development. Data processing within the CPU is extremely SEE critical since effects may show up many cycles after their actual occurrence in almost any part of the five staged pipeline and the surrounding elements. The in-order command execution itself is extremely susceptible and modifications may quickly lead to a SEFI and therefore to a system halt. Thus, occurring errors are immediately detected to prevent faulty calculated data to be written back to memory. Figure 5.4 shows a sketch of the implemented CPU. The major advantage of this design lies in the combination of chip-area-saving DMR (avoiding extensive TMR) with the static scrubbing technique. Only the most sensitive program counter part has been realized with TMR to guarantee integrity. At runtime, both doubled pipeline signals are continuously compared with each other subsequent to every calculation step. Error detection immediately leads to pipeline interruption, in which all contents are marked invalid, preventing execution of faulty commands. Both pipelines are flushed with the instruction that was in memory stage when the error occurred and CPU restarts calculation from the point the error occurred. In case an SEU has caused the error in data or clock signals, calculation continues correctly within the following calculation cycle. If an SEU within the FPGA's routing network is responsible for the miscalculation, the error is recalculated until a scrubbing cycle

---

[a]For extensive details on the radiation tolerant CPU please see [Ref. 15].

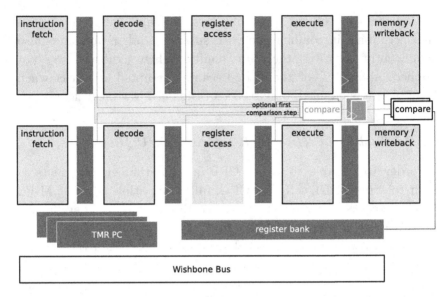

Fig. 5.4. Sketch of the fault-tolerant CPU. Two concurrent pipelines are implemented to allow detection of configuration and user bit upsets. Both pipelines share the same register bank.

repairs the defective circuits. In parallel, a signal pulse from an external scrubbing controller, indicating the refresh cycle completion state, is counted. If the CPU is detected to be irreparably stuck, the processor's registers are cleared and the CPU entirely restarts.

The CPU system has been practically tested under experimental conditions within different particle accelerator beams. The deployed DUT consists of a central Xilinx XC4VFX20 FPGA, all of the required scrubbing components as well as different readout and test interfaces. The flipchip manufactured FPGA had been directly placed within the center of the particle beam line to get comprehensible results at a maximum ionization impact. The chip had not been prepared with additional substrate or case thinning to simulate conditions found in regular application scenarios.

Test beams for the fault-tolerant CPU design used 96-Ru particles at $1.69\,\mathrm{GeV/u}$ with a flux of $1.7-5.0\cdot10^5$ ions per cm^2 each 15 seconds spill and 12-C particles at $200-500\,\mathrm{MeV/u}$ with a measured flux of $5.6\cdot10^2-2.1\cdot10^6$ ions per second at top of the FPGA. The ruthenium beam theoretically leads to 2,000 to 15,000 errors. Therefore, an unhardened version of the developed CPU with a single pipeline stopped right after 1.2 seconds (2,894 measurements), whereas the fault-tolerant double pipeline CPU runs for 15.7 seconds on average (4,930 measurements) without dropping faulty results or getting stuck. This results in a SEFI rate of 163 per hour, which is quite a high value, but still a very good one for that amount of beam intensity. The carbon beam used lower flux rates and did not raise SEUs as expected, but the higher ones caused SEFI rates up to 29 per hour (200 MeV/u, $2.1 \cdot 10^2$ at FPGA).

5.5. Conclusion

In this chapter we introduced the idea of smart module redundancy that can be used in combination with scrubbing to gain radiation tolerance in a cost-efficient way. The basic idea is to classify the FPGA design's components and to treat each component class separately. Three classes have been identified: data path, I/O and control logic. While data path and I/O mainly depend on the usage of error detection codes like CRC, the control logic is best protected by automated DMR or TMR. Special focus has been on the implementation of a radiation-tolerant CPU since this is both the most powerful and the most demanding part of the FPGA's control logic.

Two test scenarios have been used to verify the proposed approach. The first one took a GET4 Read-Out-Controller and protected it against radiation via smart module redundancy. The measurements are extremely promising and show that the resulting FPGA design is ready to be used in real world scenarios at GSI/FAIR. Moreover, it demonstrates the power of smart module redundancy. An unprotected design used 54% of the flip-flops and 36% of the LUTs. The protected design used 71% of the flip-flops and 78% of the LUTs. These are factors of 1.3 and 2.1 respectively, which is outstanding compared to the factor of 6 usually found with conventional TMR. The second test scenario focused on the implementation of a radiation tolerant CPU. Here, DMR instead of TMR has been used and proven to work.

Based on these very positive results, all our future designs will be using smart module redundancy instead of TMR in order to create FPGA designs that can withstand the high radiation at GSI/FAIR.

References

1. C. Carmichael, M. Caffrey, and A. Salazar. Correcting Single-event Upsets through Virtex Partial Configuration, Technical report, Xilinx Inc., 2000.
2. M. Caffrey *et al.* Single-Event Upsets in SRAM FPGAs, in *Proc. Military and Aerospace Applications of Programmable Logic Devices*, 2002.
3. D. Bessot and R. Velazco. Design of SEU-hardened CMOS Memory Cells: the HIT Cell, in *Proc. European Conference on Radiation and its Effects on Components and Systems*, pp. 563–570, 1993.
4. T. Calin, M. Nicolaidis, and R. Velazco. Upset Hardened Memory Design for Submicron CMOS Technology, *IEEE Transactions on Nuclear Science*, **43**(6), 2874–2878, 1996.
5. Q. Shi and G. Maki. New Design Techniques for SEU Immune Circuits, in *Proc. NASA Symposium on VLSI Design*, pp. 4.2.1–4.2.16, 2000.
6. D. Mavis. Single Event Transient Phenomena – Challenges and Solutions, in *Proc. Microelectronics Reliability and Qualification Workshop*, 2002.
7. S. Baloch, T. Arslan, and A. Stoica. Design of a Single Event Upset (SEU) Mitigation Technique for Programmable Devices, in *Proc. International Symposium on Quality Electronic Design*, pp. 330–345, 2006.
8. K. Morgan *et al.* A Comparison of TMR with Alternative Fault-Tolerant Design Techniques for FPGAs, *IEEE Transactions on Nuclear Science*, **54**(6), 2065–2072, 2007.
9. H. Deppe and H. Flemming. The GSI Event-driven TDC with 4 Channels GET4, in *Proc. Nuclear Science Symposium Conference Record*, pp. 295–298, 2009.
10. S. Manz *et al.* Radiation Mitigation Efficiency of Scrubbing on the FPGA based CBMTOF Read-out Controller, in *Proc. International Conference on Field Programmable Logic and Applications*, pp. 1–6, 2013.
11. V. Friese and C. Sturm. *CBM Progress Report*, 2011.
12. J. Gebelein and U. Kebschull. Investigation of SRAM FPGA based Hamming FSM Encoding in Beam Test, in *Proc. Radiation Effects on Components and Systems*, 2012.
13. J. Gebelein, H. Engel, and U. Kebschull. An Approach to System-wide Fault Tolerance for FPGAs, in *Proc. International Conference on Field Programmable Logic and Applications*, pp. 467–471, 2009.

14. G. Kane and J. Heinrich. *MIPS RISC Architecture (2nd Edition)*, (Prentice Hall, 1991).

15. H. Engel. *Development of a Fault Tolerant Softcore CPU for SRAM based FPGAs*, PhD thesis, Kirchoff Institute for Physics, Heidelberg University, Germany, 2009.

Chapter 6

Analysing Reconfigurable Computing Systems

Wayne Luk

Department of Computing,
Imperial College London

The distinguishing feature of a reconfigurable computing system is that the function and the interconnection of its processing elements can be changed, in some cases during run-time. However, reconfigurability is a double-edged sword: it only produces attractive results if used judiciously, since there are various overheads associated with exploiting reconfigurability in computing systems. This chapter introduces a simple approach for analysing the performance, resource usage and energy consumption of reconfigurable computing systems, and explains how it can be used in analysing some recent advances in design techniques for various applications that produce run-time reconfigurable implementations. Directions for future development of this approach are also explored.

6.1. Introduction

The exponential growth of the fabrication cost of integrated circuits makes reconfigurable computing increasingly attractive. Most industrial applications involving reconfigurable computing today, however, adopt just compile-time configuration: once a device such as an FPGA (field-programmable gate array) is configured, its configuration either does not change over its lifetime, or changes only when a new application is required. Since most reconfigurable devices can, in principle, be reconfigured as many times as needed, many researchers are curious about the conditions under which reconfigurability can be exploited effectively to improve the capabilities of an implementation, and how such implementations can be characterised to enable estimates of performance, resource usage, energy consumption and so on.

This curiosity has been driving the research of Peter Cheung and me for many years, since I joined Imperial College London and started collaborating with him. We have worked with many of our outstanding students, including Tobias Becker, Thomas Chau, Gary Chow, Joern Gause, Pete Sedcole, Shay Seng and Nabeel Shirazi, in advancing the exploitation of reconfigurable technology, especially run-time reconfigurability, for computing applications.

In the following, motivations for run-time reconfigurability will first be given. An approach for analysing reconfigurable computing systems will then be introduced. It will be followed by a description of some recent advances in design techniques. Finally, directions for future development of this approach will be explored.

6.2. Why Run-time Reconfigurability?

Many computer systems, including those for high-performance computing and for embedded applications, require designers to cope with three competing requirements: adaptability, high performance, and reduced time-to-deployment.

In a continuously evolving environment, embedded systems must adapt to changes in both function and performance. The adaptation may entail *in situ* programming to meet new protocols or upgrade to new algorithms. Similarly, accelerators for cloud computing should support a variety of functions and workloads. All these systems must deliver increasingly high performance within challenging power, size, and fault tolerance constraints, often precluding a conventional processor approach.

The requirement for both adaptability and high performance conflicts with another goal of minimising time-to-deployment. Recent hardware compilation tools are beginning to support rapid design development; they often involve application-specific design customisation to meet constraints in performance, resource usage and power consumption. Some tools also support a systematic approach to run-time adaptation.

An introduction to reconfigurable computing, including a brief description of run-time reconfigurability, is available [Ref. 1]. Run-time reconfigurability can deliver the following benefits [Refs. 2 and 3]:

- implementing a large design by time multiplexing
- accelerating demanding applications
- improving power and energy consumption
- supporting health monitoring
- enhancing reliability and fault tolerance
- speeding up the design cycle by enabling incremental development.

Next, an approach for analysing reconfigurable computing systems will be introduced. Some advances in design techniques targeting implementations with run-time reconfigurability will then be presented.

6.3. An Analysis Approach

There have been many laudable advances in techniques for performance analysis of reconfigurable computing systems [Refs. 4–9]. However, they are either designed to model various effects such as those of computation and communication in detail, or focused only on specific optimisations or on a specific application.

In contrast, we present below a simple approach for analysing performance, resource usage and energy consumption of reconfigurable computing systems. This approach is not intended to replace the existing work cited earlier, but to provide an abstract model which can be further refined to include various specific effects. Our approach is designed to focus on the key essential elements that transcend implementation details of particular technologies or systems.

Our approach involves three factors, ς, ρ and ε, capturing respectively potential improvements in completion time, resource usage and energy consumption of a reconfigurable computing system against a *conventional reference system*. In the appropriate context, the conventional reference system can be an instruction processor, an application-specific integrated circuit, or a reconfigurable processor that does not support run-time reconfiguration. This will be illustrated below.

6.3.1. *Analysis of completion time*

Let us adopt ς to denote the ratio of the completion time for a conventional reference system to the completion time for the corresponding reconfigurable computing system. The completion time for the reconfigurable computing system includes two components: the execution time and the reconfiguration time.

Let N_c denote the number of cycles of a particular application for the conventional reference system, N_e denote the number of cycles for execution for the corresponding reconfigurable computing system, and N_r denote the number of cycles for its reconfiguration. Let T_c denote the cycle time for the conventional reference system, and T_e and T_r denote respectively the cycle time for execution and for reconfiguration for the corresponding reconfigurable computing system. One can then define that

$$\varsigma = \frac{N_c T_c}{N_r T_r + N_e T_e}. \tag{6.1}$$

Clearly ς shows the speed benefit of reconfigurability relative to the conventional reference system; $\varsigma = 1$ means that reconfigurability does not deliver any speed-up. However, the reconfigurable system can still be attractive even when $\varsigma < 1$, as long as the system meets the speed requirement while having significant reduction in resource usage or in energy consumption, or both. Before we look at these cases, let us study the above equation in more detail.

First, consider the conventional reference system being an instruction processor. In that case $N_c = N_i C_i$, where N_i is the number of instructions for that application, and C_i is the average number of cycles per instruction for that application.

Second, the equation above can easily be generalised to cover designs that support multiple reconfigurations at run time:

$$\varsigma = \frac{N_c T_c}{\Sigma_j (N_{r,j} T_{r,j} + N_{e,j} T_{e,j})}. \tag{6.2}$$

Third, the effects of optimisations such as configuration prefetching can be approximated by including an additional parameter γ_j to account for the reduction in reconfiguration time due to prefetching:

$$\varsigma = \frac{N_c T_c}{\Sigma_j (\gamma_j N_{r,j} T_{r,j} + N_{e,j} T_{e,j})}. \tag{6.3}$$

Fourth, the above model can be used in exploring the trade-offs between different reconfiguration regimes. For example, it is possible that an increase in parallelism would reduce the execution time while increasing the reconfiguration time; an optimal amount of parallelism can be derived which would offer the largest reduction in completion time [Ref. 8].

Fifth, once ς is known, the effect due to Amdahl's law can be estimated: given that only a fraction β of the completion time can benefit from acceleration, the overall improvement factor δ is given by

$$\delta = \frac{\varsigma}{\varsigma(1 - \beta) + \beta}. \tag{6.4}$$

6.3.2. *Analysis of resource usage*

Let R_1 and R_2 denote the resources needed for two hardware elements that are not needed concurrently for a given application. A non-reconfigurable design would still need to have sufficient resources $R_1 + R_2$ to accommodate both elements even if only one of them is active at one time.

In contrast, a run-time reconfigurable solution would only need to have sufficient resources for $max(R_1, R_2) + R_r$, where R_r denotes the resource overhead required by the reconfigurable design to support, for example, additional storage for capturing the state between successive configurations.

The benefit ρ of reconfigurability on resource usage can then be calculated as follows:

$$\rho = \frac{R_1 + R_2}{max(R_1, R_2) + R_r}. \tag{6.5}$$

As before, this equation can easily be generalised to model designs that support multiple reconfigurations at run time:

$$\rho = \frac{\Sigma_j R_j}{max_j(R_j) + max_j(R_{r,j})}. \tag{6.6}$$

The largest gain can be obtained when all the R_j elements have the same resource usage.

For FPGAs, there are different kinds of resources such as fine-grained logic blocks, coarse-grained arithmetic and signal processing elements, and configurable memory elements. Another method is to estimate the number of transistors required for each of these resources so that an aggregate can be obtained; it would enable us to compare resource usage for other forms of processors as long as they use the same transistor technology.

6.3.3. Analysis of energy consumption

Let P_c denote the power consumption of the conventional reference system. Since energy consumption is the product of power consumption and the associated activated time, then its energy consumption will be given by $P_c N_c T_c$ where, as before, N_c and T_c denote respectively the number of cycles and the cycle time for a given application executing on the conventional reference system.

Similarly, let P_r, N_r and T_r denote respectively the power consumption, the number of cycles and the cycle time for reconfiguration of a given application on the corresponding reconfigurable computing system, and P_e, N_e and T_e denote respectively the power consumption, the number of cycles and the cycle time for its execution. The total energy consumption for the reconfigurable computing system is then given by $P_r N_r T_r + P_e N_e T_e$.

The benefit ε of reconfigurability on energy consumption can be calculated as follows:

$$\varepsilon = \frac{P_c N_c T_c}{P_r N_r T_r + P_e N_e T_e}. \tag{6.7}$$

The analysis above is deliberately kept straightforward. It will be used in the next section to illustrate how design techniques promoting reconfigurability can improve performance, resource usage or energy efficiency of implementations.

6.4. Design Techniques for Run-time Reconfigurability

This section presents three techniques that target run-time reconfigurable implementations. They are based on mixed-precision computation, multi-stage processing, and elimination of idle functions. The main consideration is to partition a design into multiple optimized configurations, which can then be placed onto the target reconfigurable engine at the appropriate instant during run time.

6.4.1. *Mixed-precision computation*

This approach involves having multiple configurations of different precisions, and it is assumed that the computation can benefit from having multiple datapaths operating concurrently. Since a datapath operating in low precision is smaller than one operating in high precision, one would include as many datapaths as possible in each configuration to maximise parallelism.

The main challenges of this approach are to find appropriate precisions that would deliver correct results with sufficient accuracy while maximising parallelism, and to schedule multiple configurations at appropriate instants.

This approach can be analysed based on the techniques introduced in the previous section. There are two steps.

(1) Let R_f denote the resources for a full-precision datapath, and N_f denote the number of such datapaths that can be accommodated on a given FPGA. Also let R_s denote the resources for a small datapath with reduced precision, and N_s denote the number of

such datapaths that can be accommodated on the same FPGA as before. Hence ideally $N_s R_s = N_f R_f \leq R_T$, where R_T denotes the total amount of resources on a device. Since $R_f > R_S$, hence $N_f < N_s$.

Let N be the total number of iterations required for this computation. When the FPGA contains only full-precision datapaths, the completion time is given by N/N_f. When the FPGA contains both full-precision datapaths and reduced-precision datapaths, if a fraction β of the iterations can take place in reduced precision, the completion time is given by $\beta N/N_s + (1 - \beta)N/N_f$. So the improvement in completion time μ due to mixed-precision computation can be expressed by

$$\mu = \frac{N_s}{\beta N_f + (1 - \beta)N_s}.$$

(2) The benefit of reconfigurability on resource usage can be derived from Eq. 6.5, assuming negligible reconfiguration overhead:

$$\rho = \frac{N_s R_s + N_f R_f}{max(N_s R_s, N_f R_f)}.$$

Let τ_f and τ_s denote respectively the execution time for the application when the system has a single full-precision datapath and when the system has a single reduced-precision datapath, and τ_r denote the time it takes to reconfigure the device, Eq. 6.1 becomes

$$\varsigma = \frac{N_c T_c}{2\tau_r + (\tau_f/N_f) + (\tau_s/N_s)}.$$

The mixed-precision approach has been applied to three application domains: Monte Carlo simulation [Ref. 10], function comparison [Ref. 11], and mathematical optimisation [Ref. 12]. While computation in full precision can also be performed in software on a general-purpose processor, if the reconfiguration overhead is low while the communication cost between the reconfigurable engine and the general-purpose processor is high, then it would be profitable to adopt run-time reconfiguration in conjunction with having both full-precision and reduced-precision computation in the reconfigurable engine. The following provides an overview of mixed-precision computation for three application domains; the performance reported below does not include run-time reconfigurability, so there is scope for further improvement.

First, Monte Carlo simulation [Ref. 10]. It involves the use of datapaths with reduced precision, and the resulting errors are corrected by auxiliary sampling. An analytical model for speed and resource usage can be developed to enable optimisation based on mixed integer geometric programming to determine the optimal reduced precision and the optimal resource allocation among the Monte Carlo datapaths and correction datapaths. Experiments show that the mixed precision approach requires up to 11% additional evaluations while less than 4% of all the evaluations are computed in full precision. The resulting designs, with the full-precision correction operations implemented in software, are up to 7.1 times faster and 3.1 times more energy efficient than baseline double-precision FPGA designs, and up to 163 times faster and 170 times more energy efficient than quad-core software designs optimised with the Intel compiler and Math Kernel Library. This approach also produces designs for pricing Asian options which are 4.6 times faster and 5.5 times more energy efficient than NVIDIA Tesla C2070 GPU implementations.

Second, function comparison [Ref. 11]. Our approach improves comparison performance by using reduced-precision datapaths while maintaining accuracy by using full-precision datapaths. The approach adopts reduced-precision datapaths for preliminary comparison, and full-precision datapaths when the accuracy for preliminary comparison is insufficient. As in Monte Carlo simulation, an analytical model for performance estimation can be developed, with optimisation based on integer linear programming used for determining the optimal precision and the optimal resource alloca-tion for each of the datapaths. The effectiveness of this approach is evaluated using a common collision detection problem with full-precision computation performed in software. Performance gains of 4 to 7.3 times are obtained over the baseline fixed-precision designs for the same FPGAs. The mixed-precision approach leads to FPGA designs which are 15.4 to 16.7 times faster than software running on multi-core CPUs with the same technology.

Third, mathematical optimisation [Ref. 12]. It involves the use of reduced precision optimisers for searching potential regions containing the global optimum, and full-precision optimisers for verifying the results. An empirical method is used in determining parameters for the mixed-precision approach. Its effectiveness is evaluated using a set of optimisation benchmarks, with full-precision optimizers implemented in software. It is found that one can locate the global optima 1.7 to 6 times faster when compared with a quad-core optimiser. The mixed-precision optimisations search up to 40.3 times more starting vectors per unit time when compared with full-precision optimisers, and only 0.7% to 2.7% of these searches are refined using full-precision optimisers. This approach allows us to accelerate problems with more complicated functions or to solve problems involving higher dimensions.

6.4.2. *Multi-stage processing*

If a computation can be partitioned into multiple stages such that each stage requires different amounts of resources, then each stage can be captured in a configuration such that successive stages can be realised by run-time reconfiguration. Two examples will be used to illustrate this approach.

First, short-read sequence alignment [Ref. 7]. This computation involves determining the positions of millions of short reads relative to a known reference genetic sequence. Our approach consists of two key components: an exact string matcher for the bulk of the alignment process, and an approximate string matcher for the remaining cases. Based on techniques similar to those in Sec. 6.3, interesting regions of the design space, including homogeneous, heterogeneous and run-time reconfigurable designs, can be characterised to provide performance estimations of the corresponding designs. An implementation of a run-time reconfigurable architecture targeting a single FPGA can be up to 293 times faster than running the BWA algorithm on an Intel X5650 CPU, and 134 times faster than running the SOAP3 program on an NVIDIA GTX 580 GPU.

Second, adaptive particle filters [Ref. 13]. This is a statistical method for dealing with dynamic systems having non-linear and non-Gaussian properties. It has been applied to real-time applications including object tracking, robot localisation, and air traffic control. However, the method involves a large number of particles resulting in long execution times, which limits its application in real-time systems. An approach has been developed to adapt the number of particles dynamically and to utilise run-time reconfigurability of the FPGA for reduced power and energy consumption. The performance of designs based on this approach can be analysed using techniques similar to those in Sec. 6.3. Implementations for simultaneous mobile robot localisation and people tracking show that adaptive particle filters can reduce up to 99% of computation time. Using run-time reconfiguration, 34% reduction in idle power and 26–34% reduction of system energy can be achieved. Moreover, the proposed system is up to 7.39 times faster and 3.65 times more energy efficient than the Intel Xeon X5650 CPU with 12 threads, and 1.3 times faster and 2.13 times more energy efficient than an NVIDIA Tesla C2070 GPU.

6.4.3. *Eliminating idle functions*

Idle functions have a detrimental effect on performance, area and power consumption. A design approach has been developed to automatically identify and exploit run-time reconfiguration opportunities [Ref. 14], while optimising resource utilisation by eliminating idle functions. The approach is based on a hierarchical graph structure which enables run-time reconfigurable designs to be synthesised in three steps: function analysis, configuration organisation, and run-time solution generation.

The performance of designs based on this approach can be analysed in a similar way as those in Sec. 6.4.1. In this case, more datapaths can be accommodated in different configurations by eliminating idle functions than one without run-time reconfiguration and having functions that may be idle during run-time due to, for example, data dependence constraints.

Three applications, targeting barrier option pricing, particle filter, and reverse time migration, are used in evaluating the proposed approach. Our reconfigurable implementations are 1.31 to 2.19 times faster than optimised static designs, and are up to 28.8 times faster than optimised CPU reference designs and 1.55 times faster than optimised GPU designs.

6.5. Future Development

Run-time reconfigurability has shown good promise in producing efficient implementations. However, much research remains to be conducted before run-time reconfigurable design can be adopted as a mainstream vehicle for realising computer systems.

First, it would be beneficial to develop further the theoretical foundations of run-time reconfigurability; the approach in Sec. 6.3 can be regarded as preliminary work in this direction. Such theories would, for example, provide an analytical treatment of optimality which would be useful for developing design techniques. An attempt has been made for providing such analytical treatment that models the combined effects of parallelisation and reconfiguration on performance [Ref. 8] and on energy efficiency [Ref. 9], which reveals that there is often an optimal design with the highest performance or with the highest energy efficiency. We hope to extend and generalise the above work and related techniques to provide the basis for tools that generate optimised designs.

Second, while there is an increasing number of tools targeting run-time reconfigurable designs, developing and optimising such designs is still much more difficult and tedious than, for example, software optimisation. Promising approaches addressing this issue include aspect-oriented techniques [Ref. 15] that promote separation of concerns and design re-use; incremental design [Ref. 3] that helps to speed up design implementation; and the use of domain-specific descriptions [Ref. 16] to raise the level of abstraction for descriptions that target application builders.

Third, there appears significant synergy between reconfigurable computing and machine learning. In particular, machine learning techniques have been used in speeding up reconfigurable design [Ref. 17] while reconfigurable design has also been used in speeding up machine learning [Ref. 18]. We hope to understand more deeply the connections between these two areas, and how such synergy can contribute to advances in both areas.

6.6. Summary

Reconfigurable computing in general, and run-time reconfigurability in particular, have come a long way in the last 20 years. Exciting advances have recently been made, resulting in some of the fastest and the most energy-efficient designs for various applications. We shall accelerate the progress of our research on various aspects of reconfigurable computing and to extend its influence, so that we would come closer to realising the vision of Peter Cheung for the next 20 years: that reconfigurability will be found in all integrated circuits, not just FPGAs as we know today.

Acknowledgements

Many thanks to James Arram, Tobias Becker, Thomas Chau, Gary Chow, Andreas Fijeland, Qiwei Jin, Maciej Kurek, Philip Leong, Qiang Liu, Stephen Muggleton, Xinyu Niu, Henry Styles, David Thomas, Brittle Tsoi and Steve Wilton for contributing to this paper. I am grateful to Peter Cheung for his advice, enthusiasm and friendship over many years. This research is supported in part by the European Union Seventh Framework Programme under grant agreement number 257906, 287804 and 318521, by the UK EPSRC, by the Maxeler University Programme, by Altera, and by Xilinx.

References

1. T. Todman *et al.* Reconfigurable Computing: Architectures and Design Methods, *IEE Computer and Digital Techniques*, 152(2), 193–207, 2005.

2. P. Lysaght *et al.* Enhanced Architectures, Design Methodologies and CAD Tools for Dynamic Reconfiguration of Xilinx FPGAs, in *Proc. International Conference for Field Programmable Logic and Applications*, pp. 1–6, 2006.

3. T. Frangieh *et al.* PATIS: Using Partial Configuration to Improve Static FPGA Design Productivity, in *Proc. IPDPS Workshop*, pp. 1–8, 2006.

4. E. El-Araby, I. Gonzalez, and T. El-Ghazawi. Exploiting Partial Runtime Reconfiguration for High-performance Reconfigurable Computing, *ACM Transactions on Reconfigurable Technology and Systems*, 1(4), 21:1–21.23, 2009.

5. E. Holland, K. Nagarajan, and A. George. RAT: RC Amenability Test for Rapid Performance Prediction, *ACM Transactions on Reconfigurable Technology and Systems*, 1(4), 22:1–22:31, 2009.

6. K. Papadimitriou, A. Dollas, and S. Hauck. Performance of Partial Reconfiguration in FPGA Systems: A Survey and a Cost Model, *ACM Transactions on Reconfigurable Technology and Systems*, 4(4), 36:1–36:24, 2011.

7. J. Arram *et al.* Reconfigurable Acceleration of Short Read Mapping, in *Proc. International Symposium on Field-Programmable Custom Computing Machines*, pp. 210–217, 2013.

8. T. Becker, W. Luk, and P. Cheung. Parametric Design for Reconfigurable Softwaredefined Radio, in *Proc. International Symposium on Applied Reconfigurable Computing*, pp. 15–25, 2009.

9. T. Becker, W. Luk, and P. Cheung. Energy-aware Optimisation for Run-time Reconfiguration, in *Proc. International Symposium on Field-Programmable Custom Computing Machines*, pp. 55–62, 2010.

10. G. Chow *et al.* A Mixed Precision Monte Carlo Methodology for Reconfigurable Accelerator Systems, in *Proc. International Symposium on Field Programmable Gate Arrays*, pp. 57–66, 2012.

11. G. Chow *et al.* Mixed Precision Processing in Reconfigurable Systems, in *Proc. International Symposium on Field-Programmable Custom Computing Machines*, pp. 17–24, 2011.

12. G. Chow, W. Luk, and P. Leong. A Mixed Precision Methodology for Mathematical Optimisation, in *Proc. International Symposium on Field-Programmable Custom Computing Machines*, pp. 33–36, 2011.

13. T. Chau *et al.* Heterogeneous Reconfigurable System for Adaptive Particle Filters in Real-time Applications, in *Proc. International Symposium on Applied Reconfigurable Computing*, pp. 1–12, 2013.

14. X. Niu *et al.* Automating Elimination of Idle Functions by Run-time Reconfiguration, in *Proc. International Symposium on Field-Programmable Custom Computing Machines*, pp. 97–104, 2013.

15. J. Cardoso *et al.* Specifying Compiler Strategies for FPGA-based Systems, in *Proc. International Symposium on Field-Programmable Custom Computing Machines*, pp. 192–199, 2012.

16. D. Thomas and W. Luk. A Domain Specific Language for Reconfigurable Path-based Monte Carlo Siumlations, in *Proc. International Conference on Field-Programmable Technologies*, pp. 97–104, 2007.

17. M. Kurek, T. Becker, and W. Luk. Parametric Optimisation of Reconfigurable Designs using Machine Learning, in *Proc. International Symposium on Applied Reconfigurable Computing*, pp. 134–145, 2013.
18. A. Fidjeland, W. Luk, and S. Muggleton. A Customisable Multiprocessor for Application-optimised Inductive Logic Programming, in *Proc. Visions of Computer Science — BCS International Academic Conference*, 2008.

Chapter 7

Custom Computing or Vector Processing?

Simon W. Moore, Paul J. Fox, A. Theodore Markettos
and Matthew Naylor

Computer Laboratory, University of Cambridge

FPGAs are famously good for constructing custom arithmetic pipelines for video processing and other data intensive tasks, but building these custom pipelines is time-consuming. Moreover, large computation tasks typically require large data-sets and managing the memory bottleneck is crucial. This chapter demonstrates that vector processing can not only deliver high computational performance, but that the memory bottleneck can be effectively managed too. Whilst more general purpose vector compute is unlikely to deliver the full performance of a custom pipeline, we demonstrate that for a non-trivial neural computation case study we can get within a factor of two of a custom solution.

7.1. Introduction

There is a great deal of research on efficiently mapping algorithms onto FPGAs that produces custom computation pipelines, which aim to exploit the massively parallel computation resources available on today's FPGAs. Examples from Professor Peter Cheung's work include video processing [Ref. 1], wavelets [Ref. 2], elliptic curve cryptography [Ref. 3] and prime number validation [Ref. 4]. Constructing complex custom pipelines is time consuming, though suitable abstractions such as C-to-gates improve productivity, but often at the expense of performance. Vector processing can be an attractive alternative to C-to-gates [Ref. 5], yielding good performance with the convenience of software programming and debugging.

Applications demanding massive compute often require large data-sets, and this can lead to streaming data from memory external to a FPGA becoming a bottleneck for a broad class of applications whose pinnacle of performance is reached when external memory bandwidth is saturated with useful data transfer, not when the FPGA compute resources are maximally used [Ref. 6]. This is known as the "memory wall", and is an increasing problem for both ASICs and FPGAs [Ref. 7], where compute resources are more plentiful than external memory bandwidth. In this chapter, we focus on this class of application with an in-depth case study of neural computation which has demanding data interdependencies and large data-sets that need to be held in external memory.

Our previous work on custom neural computation pipelines for FPGAs resulted in a high-performance design described using Bluespec HDL [Ref. 8], capable of efficiently streaming neuron and synapse parameters from DDR2 memory to achieve real-time performance with 64 k neurons and 64 M synapses per Altera Stratix IV 230 FPGA [Ref. 9].

This custom pipeline implementation took around three man-years to complete, and resulted in us having a deep understanding of the Izhikevich spiking neuron model. While highly parameterised, this implementation is still rather inflexible (e.g. if the neuron model has to be changed) and is therefore of less utility to neuroscientists (our prospective customers) than existing software-based neural computation systems. However, this work did identify that given some modest parallel compute, external memory bandwidth becomes a performance bottleneck for this application. As a consequence, we explore vector processing for neural computation, with a particular focus on making efficient use of external memory bandwidth using burst transfers.

Section 10.2 presents BlueVec, a vector co-processor for an Altera NIOS II. Section 7.3 uses neural computation as a case study to compare and contrast custom computing, vector processing, and multi-core implementations, and the results of this case study are given in Sec. 10.5. Section 10.7 provides conclusions and considers their implications on future research directions.

7.2. BlueVec Architecture

Recent work at the University of British Colombia has led to a series of soft vector processors [Refs. 7, 10–12] allowing the rapid development of high-performance, low-area, program accelerators on FPGAs. Of these, VIPERS [Ref. 7] is perhaps the most interesting for neural computation applications: it supports *lane-local memories* that can be addressed independently and in parallel using a vector of addresses and, as we discover, this feature is ideal for parallel distribution of synaptic updates to neurons scattered throughout memory. Unfortunately VIPERS lacks an important feature for these applications: *burst memory access* for high-performance streaming of data from external memory. While the successors to VIPERS [Refs. 10–12] have made great progress towards optimising external memory bandwidth efficiency, they all omit lane-local memories. Therefore we have developed our own soft vector processor — *BlueVec* — to meet both requirements.

7.2.1. Vector width

BlueVec [Ref. 13] is a minimalist vector co-processor — written in around 1,000 lines of Bluespec HDL — with two external interfaces: (1) a *custom instruction slave* interface for connection to a NIOS II, and (2) a *memory mapped master* interface for connection to external memory. Assuming a NIOS II clock frequency of 200 MHz, and a DDR2 external memory transferring 64 bits of data on both edges of a 400 MHz clock, the maximum data transfer rate between processor and memory is 256 bits per NIOS II clock cycle. This motivates processing vectors of 256 bits per clock cycle, which can be treated as either:

- 8 × 32 bit words (W instructions) *or*
- 16 × 16 bit half-words (H instructions) *or*
- 32 × 8 bit bytes (B instructions).

7.2.2. *Register file and instruction set*

Given that a NIOS II custom instruction is defined as containing three 5-bit register operands, the obvious design choice for BlueVec is a three-operand vector instruction set with a 32-element register file. We take this option, but there are alternatives, e.g. a large scratchpad in place of a register file with support for long vectors, which would allow a greater number of vector lanes and reduce loop overhead [Ref. 11].

An illustrative portion of the BlueVec instruction set is shown in Fig. 7.1. Note the use of v and s prefixes to distinguish BlueVec vector registers and NIOS II scalar registers respectively. All BlueVec instructions are implemented as C macros which expand to inline assembly code. Hence any valid C expression or variable can be used in place of a scalar register, but vector registers must be constants in the range v0...v31. The following sections discuss parts of the BlueVec instruction set in more detail.

7.2.3. *External memory*

Vectors can be loaded from external memory using the instruction

$$\text{Load}(vDest,\ sAddr,\ burstLength).$$

$\text{Mul[B|H|W]}(vDest,vSrcA,vSrcB) = vDest[i] \leftarrow vSrcA[i] \times vSrcB[i]$

$\text{Cmp[B|H|W]}(vDest,vSrçA,vSrcB) = vDest[i] \leftarrow \textbf{if } vSrcA[i] \leq vSrcB[i] \textbf{ then } 1 \textbf{ else } 0$

$\text{Cond[B|H|W]}(vDest,vElse,vCond) = vDest[i] \leftarrow \textbf{if } vCond[i] \textbf{ then } vDest[i] \textbf{ else } vElse[i]$

$\text{Index[B|H|W]}(sDest,vSrc,sIndex) = sDest \leftarrow vSrc[sIndex]$

$\text{Set[B|H|W]}(vDest,sMask,sSrc) = vDest[i] \leftarrow \textbf{if } sMask\ \&\ 2^i \textbf{ then } sSrc \textbf{ else } vDest[i]$

$\text{LoadLocalH}(vDest,vAddr) = vDest[i] \leftarrow LOCAL_i[vAddr[i]]$

$\text{Load}(vDest,sAddr,burstLength) = vDest,\dots,(vDest + burstLength - 1) \leftarrow$
$$MEM[sAddr],\dots,MEM[sAddr + burstLength - 1]$$

Fig. 7.1.   An illustrative portion of the BlueVec [Ref. 13] instruction set. Register names prefixed with v denote 256-bit vectors and those prefixed with s denote 32-bit scalars. Index i ranges from 0 to 31, 15, and 7 for byte (B), half-word (H), and full-word (W) instructions respectively. $LOCAL_i$ denotes lane local memory i, and MEM denotes external memory with a 256-bit data bus.

When executed, a *burstLength*-element sequence of 256 bit vectors beginning at address *sAddr* is read into registers

$$vDest, (vDest + 1), \ldots, (vDest + burstLength - 1).$$

However, the register file is *not modified* until a `Commit` instruction is issued. This allows the latent `Load` instruction to be a non-blocking operation that can be issued well before its result is actually needed, where need is signified by a blocking `Commit`. For example, data for the next iteration of a loop can be fetched while data for the current iteration is processed. In principle, `Commit` instructions can be inferred in hardware using register scoreboarding, but we have opted for an explicit design to keep the hardware simple.

The corresponding store instruction is already a non-blocking operation and, at the time of writing, does not support bursts:

$$\texttt{Store}(vSrc, sAddr).$$

7.2.4. *Lane-local memories*

Each half-word vector lane has its own local Block RAM giving a local memory that is accessible by a vector of 16 addresses. The instruction

$$\texttt{LoadLocalH}(vDest, vAddr)$$

loads $LOCAL_i[vAddr[i]]$ into $vDest[i]$ for each of the lane local memories $LOCAL_i$ where $i \in \{0 \ldots 15\}$. The size of each $LOCAL_i$ is a BlueVec design parameter that can be altered on a per-application basis. The corresponding store instruction is

$$\texttt{StoreLocalH}(vSrc, vAddr).$$

Only half-word versions of these instructions are supported since 16-bit addresses are more appropriate for medium-sized block RAMs than 8 or 32 bits. Full-word variants can be coded by duplicating each 16-bit address to form a 32-bit address.

7.2.5. *Pipelining*

In order to achieve a clock frequency above 200 MHz, i.e. not inhibit the NIOS II clock frequency, BlueVec uses a 3 stage pipeline:

- F: operand fetch (from register file)
- E: execute instruction
- W: writeback result (to register file).

Most instructions execute in a single cycle and provide a result that can be used immediately. This is made possible by *register forwarding*: the result and destination register of stages E and W are inspected by stage F and used to override, if necessary, the values fetched from the register file.

There are three instructions which do not fully complete in a single cycle:

- `LoadLocalH` completes in single cycle but the caller must wait one further cycle before reading the result; register forwarding is not possible at stage E since the output of block RAM is not yet available. Waiting can be achieved using `NoOp` or any other instruction that does not read the destination register.
- `Mul` completes in two cycles but the caller must wait two further cycles before reading the result. This is due to a 3 cycle latency on FPGA multiplier blocks clocked at over 200 MHz. A deeper pipeline could alleviate this delay.
- `Index` takes 3 cycles to complete since it must pass through the whole pipeline before a result can be returned to the NIOS II.

7.2.6. *Record/playback*

We observed that the NIOS II is unable to issue custom vector instructions at the maximum possible rate of one per cycle. As a workaround, we introduced a record/playback facility which allows sequences of instructions to be written to a local instruction memory inside BlueVec, and played back at the maximum rate by issuing a single instruction.

To illustrate this mechanism,

```
Record(start);
 // Sequence of vector instructions
Record(end);
```

records an instruction sequence that can be played back by calling:

```
Playback(start,end).
```

7.3. Neurocomputing Case Study

As a case study we compare our custom neural computation pipeline [Ref. 9] for the Izhikevich spiking neuron model [Ref. 14] to an implementation of the same algorithm using BlueVec. External memory bandwidth is critically important since the FPGA has insufficient BRAM to hold all of the neural parameters. Figure 7.2 illustrates the data that needs to be stored:

(1) The neuron firing rule equation parameters are stored sequentially in memory.
(2) When the rule fires a pointer is used to deference a list of fan-out tuples containing (delay, pointer).
(3) After the appropriate delay a long list of (neuron ID, weight) pairs are read which need to be processed by summing the weight for the appropriate neuron's *I-value*.

These data structures are intended to optimise the size of burst reads to external memory made by each phase of the algorithm,

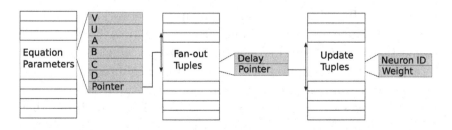

Fig. 7.2. Layout of data in external memory.

and hence bandwidth usage efficiency. Further details can be found in [Ref. 15].

The custom neural computation pipeline uses a separate computation unit for each of the three phases of the neural computation algorithm, each performing burst reads to external memory, and with pointers being passed between them using FIFO communication channels. While the use of Bluespec SystemVerilog [Ref. 8] rather than conventional Verilog, improved productivity, this implementation still required three man-years to complete.

The BlueVec implementation of the Izhikevich spiking neural model uses the same data structures in external memory as used by the custom pipeline implementation. Its implementation required one man-week to complete in addition to the time required to implement the BlueVec architecture itself. While productivity for this implementation was aided by reuse of the design effort needed for data structures and common components such as DDR2 memory controllers, this still represents a striking difference.

To provide a more specific example of the reduction in design effort afforded by using a a BlueVec vector coprocessor rather than a custom computation pipeline, we will focus on the application of synaptic updates phase of the Izhikevich spiking neuron model (*I-value accumulation*). Details of the full algorithm can be found in [Ref. 15].

7.3.1. *I-value accumulation*

A spiking neural network consists of neurons connected by synaptic connections. In a typical biologically plausible network, each neuron has synaptic connections to around 10^3 other neurons. Each synaptic connection has an associated *delay* and a *weight*, which signifies the strength of the connection. Collectively the combination of target neuron, delay and weight are known as a *synaptic update*. When a neuron spikes each synaptic update needs to be delayed and then summed with other synaptic updates targeted at the same neuron to produce a total input current (termed as an *I-value*) for each neuron. We refer to this process of summing synaptic updates as *I-value accumulation.*

If the *I*-value of neuron n is denoted ivalues[n], and its target connections and associated weights are stored in arrays targets and weights respectively, then the *I*-value accumulation process required if neuron n spikes is defined by the following loop:

```
for (i = 0; i < numTargets; i++)
    ivalues[targets[i]] += weights[i];.
```

In both the custom pipeline and BlueVec implementations, each array has elements which are 16 bits in size. Since the number of *I*-values is equal to the number of neurons, there is ample capacity for ivalues to be stored in on-FPGA Block RAM, which has a total size of 2 MB on a Stratix IV 230. However, as the number of synaptic connections is typically $10^3 \times$ the number of neurons, the targets and weights arrays for each neuron must be stored in external memory.

7.3.2. *Implementation A: Custom pipeline*

Assuming the targets and weights arrays are interleaved in memory to give an array of (target, weight) pairs called *update tuples*, our custom pipeline implementation of the *I*-value accumulation loop (known as the accumulator block in previous work) is shown in Fig. 7.3. The on-FPGA Block RAM used to store the *I*-values is partitioned into eight banks, since eight update tuples can be obtained (in a single 256-bit DDR2 memory transfer) per clock cycle when efficient burst reads are used. Each bank is surrounded by a pipeline which processes update tuples. Each update tuple in an incoming word is then allocated to the bank that holds the *I*-value for the target neuron, with FIFO queues and arbiters used to allow multiple update tuples in the same 256-bit word to target the same bank.

While the FIFO queues do provide some tolerance of uneven load between banks, in practice it was found that highest performance was achieved when update tuples were arranged in 256-bit words such that they are effectively statically scheduled, with the update

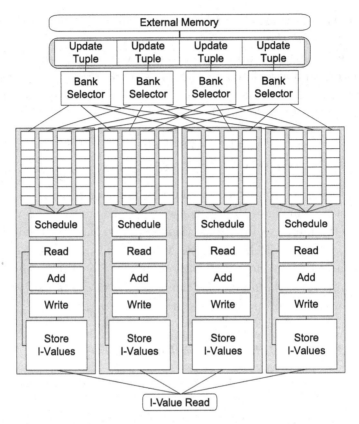

Fig. 7.3.   Custom pipeline implementation of the I-value accumulation phase. Four banks are shown for clarity — there are actually eight banks.

tuple in position x of a 256-bit word always targeting a neuron whose identifier modulo 8 is equal to x (or else being empty, denoted by zero weight).

As a result of this static scheduling, the complex accumulator block effectively becomes a vector of independent blocks, and hence the function it performs is amenable to implementation using the BlueVec vector processor.

7.3.3.   Implementation B: Vector processing

Figure 7.4 shows a BlueVec implementation of I-value accumulation. While this vectorised loop gives good speed-up over the simple

```
// Read 16 targets and weights into
// vectors registers 8 and 16
Load(v8, targets, 1);
Load(v16, weights, 1);
for (i = 0; i < numTargets; i+=16) {
    Commit;

    // Pre-fetch values for next
    // iteration in background
    Load(v8, targets+i+16, 1);
    Load(v16, weights+i+16, 1);

    // Update 16 I-values
    LoadLocalH(v0, v8);
    NoOp;
    AddH(v0, v0, v16);
    StoreLocalH(v0, v8);
}
```

Fig. 7.4. BlueVec implementation of the *I*-value accumulation phase (without bursts and record/playback). *I*-values are stored in lane-local memories.

scalar loop, it is markedly improved by changing the *burstLength* argument of each **Load** instruction from 1 to 8. Consequently, the loop increment changes from 16 to 128 and each of the four instructions at the end of the loop is performed eight times as follows:

```
LoadLocalH(v0, v8); NoOp;
AddH(v0, v0, v16); StoreLocalH(v0, v8);
...
LoadLocalH(v0, v15); NoOp;
AddH(v0, v0, v23); StoreLocalH(v0, v15);.
```

This results in a very long sequence of vector instructions that can be efficiently issued at a rate of one per clock cycle using the record/playback feature.

7.4. Results

We now discuss the performance and productivity of our custom computing and vector processing approaches to neural computation. Each approach was implemented on a Terasic DE4 evaluation board with a Stratix IV 230 FPGA, using a single DDR2 external memory bank.

7.4.1. *Custom computing performance*

Figure 7.5 illustrates the neural spike pattern for a synthetic neural network with biologically plausible numbers of synapses and firing rate running on the full custom pipeline system. The aim was to achieve real-time performance which is demonstrated by Fig. 7.6 which plots the number of neurons spiking in each millisecond time interval (in top diagram), the number of clock cycles needed to complete the resulting work (in bottom diagram), and the time bound (horizontal line in bottom diagram) given that the system is running at 200 MHz and the neural model is running at a sampling interval of 1 ms. Although not plotted, we can report that the DDR2 bandwidth utilisation is around 75% for this model.

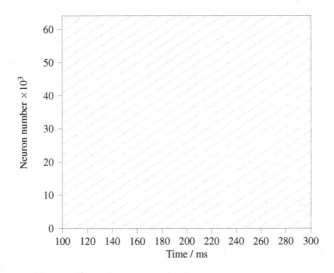

Fig. 7.5.   Spike pattern for a computation of 64 k neurons.

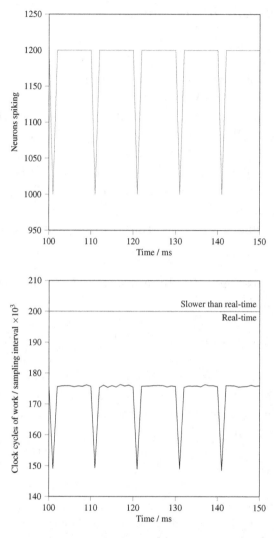

Fig. 7.6. Neurons spiking per sampling interval and clock cycles of work per sampling interval for computation of 64k neurons.

We can see that real-time performance is met, so 1 s of neural model time takes 1 s to complete. We'll see for the vector processor versions that 1 s of neural model time takes longer to complete due to inefficiencies introduced using a more general-purpose software-programmable approach.

7.4.2. Single-core performance

Table 7.1 shows the times taken by our scalar (without BlueVec) and vector (with BlueVec) Izhikevich neural computation systems to compute 1s of neural activity in a benchmark network consisting of 64 k neurons with 64 M synaptic connections [Ref. 15]. Note that the scalar version has been optimised for performance: I-values are stored in a large Block RAM, and the NIOS II has a 4 kB data cache with 256-bit cache lines that can be filled by single DDR2 memory transfers. Our custom pipeline implementation operates in real-time, around 80× faster than the scalar version, but only 4× faster than the vector version.

The performance profiles in Table 7.1 are split into three phases. The time for the I-value accumulation stage (discussed in Sec. 7.3) is reduced by 40× using vector processing. The neuron update and spike delay phases have not been discussed here, but details can be found in [Ref. 15].

We observed that DDR2 bandwidth utilisation for the single-threaded vector version was 16%. The fact that a single BlueVec is not able to saturate memory bandwidth is a consequence of an imbalance between memory access and compute. For example, while the states of 16 neurons can be fetched in 6 memory transfers, 44 instructions are required to process them. There is scope for improvement by increasing the number of operational units in the vector processor. But in the meantime, we explore the use of multiple cores to saturate memory bandwidth.

Table 7.1.   Run time and % of total for phases of Izhikevich neuron model with and without a BlueVec co-processor.

	NIOS II		+ BlueVec	
	Time/s	%	Time/s	%
I-values	57.2	72	1.4	36
Neuron updates	17.2	22	1.7	44
Spike delay buffer	1.6	2	0.5	13
Total	79.0		3.9	

7.4.3. *Multi-core performance*

Neural computation is a highly parallel task, and our benchmark neural network is easily split into smaller networks that can be processed in parallel with negligible communication. Table 7.2 shows the performance improvement obtained with multiple NIOS II and BlueVec cores accessing shared DDR2 memory and distributed Block RAMs for stack and instruction memory. The cores are connected in a star network, with a master core connected bidirectionally to all slave cores. Notably, the quad-core BlueVec configuration gives performance that is well within a factor of two of our custom pipeline. Interestingly, the two have almost identical logic utilisation.

Both the scalar and vector versions show good scaling to multiple cores. However, the sheer number of scalar cores required to keep up raises several concerns:

- The number of scalar cores that would be required to match the quad-core BlueVec implementation is well beyond the capacity of a single FPGA.
- As the number of cores increases, memory accesses become increasingly fragmented, and the performance of DDR2 memory drops.
- Inter-processor communication requirements grow as a neural network is divided into increasingly small chunks. In general, a

Table 7.2. Run time and logic and bandwidth utilisation for a multi-threaded Izhikevich neuron simulator with varying number of cores and vector co-processors.

NIOS II cores	BlueVec cores	Time (s)	Logic (%)	Bandwidth DDR2 (%)
2	0	40.2	14	1.9
4	0	20.9	19	3.8
8	0	10.8	30	7.5
16	0	6.1	53	13.1
2	2	2.2	26	30.5
4	4	1.3	49	51.0

simple star network will not scale, and more logic will be needed to efficiently connect the multitude of cores together.

7.4.4. *Productivity*

Table 7.3 shows the number of lines of code in our neural computation systems. The almost 3 k lines needed for the custom pipeline implementation is particularly striking, indicating the extra level of detail that a hardware designer must express. In fact, this line count would be even higher if it were to include general-purpose libraries developed in-house. Hardware development cycles can be slow for other reasons too, such as long synthesis times, trial-and-error refinements needed to meet tight timing constraints, and a lack of convenient I/O mechanisms for debugging.

The convenience of a software-based approach has allowed us to develop other efficient neural computation systems in a very short time. Figure 7.7 shows a screenshot of our leaky integrate-and-fire (LIF) system running the Nengo [Ref. 16] model for digit recognition on a DE4 FPGA with touch-screen. Using a single vector processor, we were able to achieve a 20× speed-up over a scalar implementation with just two days work. We do not believe that implementing a custom pipeline LIF system in this time-frame is possible, even with re-use of components from the Izhikevich system.

Table 7.3. Lines of code for various implementations of Izhikevich and LIF neuron models.

Model	Implementation	Lines
Izhikevich	Single-threaded	193
	Single-threaded and Vectorised	417
	Multi-threaded	265
	Multi-threaded and Vectorised	511
	Custom pipeline	2700
LIF	Single-threaded	324
	Single-threaded and Vectorised	496

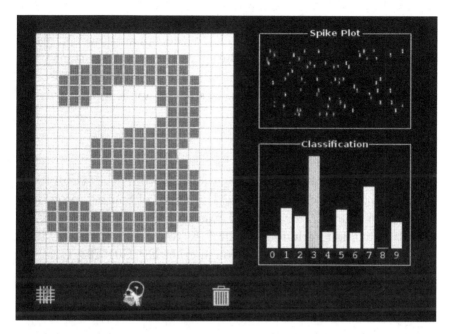

Fig. 7.7. An implementation of the Nengo digit-recognition model on a DE4 FPGA board with touch-screen.

7.5. Conclusion

We have demonstrated that a quad-core BlueVec (NIOS II + vector) machine has performance that falls within a factor of two of the custom pipeline for our neural computation case study, and has similar logic usage. The Bluespec code for BlueVec together with vectorised C code proved to be much more compact and easy to develop than the custom pipeline despite using a high-level language (Bluespec).

Vector processing also allows the memory bottleneck to be managed through efficient use of burst access to DDR2/DDR3 memory. Given that the memory bandwidth can easily become the bottleneck, inefficiency in the compute vector compute structures on FPGA becomes mostly irrelevant. Perhaps it is not too surprising that a vector multiprocessor can achieve excellent performance on FPGA given that GPUs (which are also vector multiprocessors) are highly competitive [Ref. 17]. FPGAs do, however, offer the flexibility to tailor the vector unit to the application, e.g. using custom arithmetic operations. But perhaps more importantly for massively parallel systems, FPGAs offer the ability to customise the inter-chip communication network to suit the application.

Acknowledgements

Many thanks are due to our colleagues Prof. S.B. Furber (University of Manchester), Prof. A.D. Brown (University of Southampton) and Mr. Steven Marsh (University of Cambridge) for collaborating on neural simulation techniques. The UK research council, EPSRC, provided much of the funding through grant EP/G015783/1.

References

1. N.P. Sedcole *et al.* Run-Time Integration of Reconfigurable Video Processing Systems, *IEEE Transactions on VLSI Systems*, 15(9), 1003–1016, 2007.
2. M.E. Angelopoulou *et al.* A Comparison of 2-D Discrete Wavelet Transform Computation Schedules on FPGAs, in *Proc. International Conference on FieldProgrammable Technology*, pp. 181–188, 2006.
3. R.C.C. Cheung, W. Luk, and P.Y.K. Cheung. Reconfigurable Elliptic Curve Cryptosystems on a Chip, in *Proc. Design, Automation & Test in Europe Conference & Exposition*, pp. 24–29, 2005.
4. R.C.C. Cheung *et al.* A Scalable Hardware Architecture for Prime Number Validation, in *Proc. International Conference on Field-Programmable Technology*, pp. 177–184, 2004.
5. J. Yu *et al.* Vector Processing As a Soft Processor Accelerator, *ACM Transactions on Reconfigurable Technology and Systems*, 2(2), 12:1–12:34, 2009.
6. Q. Liu *et al.* Compiling C-like Languages to FPGA Hardware: Some Novel Approaches Targeting Data Memory Organization, *The Computer Journal*, 54(1), 1–10, 2011.
7. S.A. McKee. Reflections on the Memory Wall, in *Proc. Conference on Computing Frontiers*, p. 162, 2004.
8. R.S. Nikhil and K.R. Czeck. *BSV by Example*, CreateSpace, 2010.
9. S.W. Moore *et al.* Bluehive — A Field-Programable Custom Computing Machine for Extreme-Scale Real-Time Neural Network Simulation, in *Proc. International Symposium on Field-Programmable Custom Computing Machines*, pp. 133–140, 2012.

10. P. Yiannacouras, J.G. Steffan, and J. Rose. VESPA: Portable, Scalable, and Flexible FPGA-based Vector Processors, in *Proc. International Conference on Compilers, Architectures and Synthesis for Embedded Systems*, pp. 61–70, 2008.

11. C.H. Chou *et al.* VEGAS: Soft Vector Processor with Scratchpad Memory, in *Proc. International Symposium on Field Programmable Gate Arrays*, pp. 15–24, 2011.

12. A. Severance and G. Lemieux. VENICE: A Compact Vector Processor for FPGA Applications, in *Proc. International Conference on Field-Programmable Technology*, pp. 261–268, 2012.

13. M. Naylor *et al.* Managing the FPGA Memory Wall: Custom Computing or Vector Processing?, in *Proc. International Conference on Field Programmable Logic and Applications*, 2013.

14. E.M. Izhikevich. Simple Model of Spiking Neurons, *IEEE Transactions on Neural Networks*, 14(6), 1569–1572, 2003.

15. P.J. Fox. *Massively Parallel Neural Computation. Technical Report UCAM-CL-TR-830*, University of Cambridge, Computer Laboratory, 2013. Available at: http://www.cl.cam.ac.uk/techreports/UCAM-CL-TR-830.pdf.

16. C. Eliasmith *et al.* A Large-Scale Model of the Functioning Brain, *Science*, 338(6111), 1202–1205, 2012.

17. B. Cope *et al.* Performance Comparison of Graphics Processors to Reconfigurable Logic: A Case Study, *IEEE Transactions on Computers*, 59(4), 433–448, 2010.

Chapter 8

Maximum Performance Computing
with Dataflow Technology

Michael Munday*, Oliver Pell*, Oskar Mencer*,†
and Michael J. Flynn*,‡

*Maxeler Technologies
†Imperial College London
‡Stanford University

Reconfigurable computers, generally based upon field programmable gate array (FPGA) technology, have been used successfully
as a platform for performance critical applications in a variety
of industries. Applications targeted at reconfigurable computers
can exploit their fine-grained parallelism, predictable low latency
performance and very high data throughput per watt. Traditional
techniques for designing configurations are, however, generally
considered time-consuming and cumbersome and this has limited
commercial reconfigurable computer usage. To solve this problem
Maxeler Technologies Ltd, working closely with Imperial College
London, have developed powerful new tools and hardware based
on the dataflow computing paradigm. In this chapter we explore
these tools and provide examples of how they have enabled
the development of a number of high performance commercial
applications.

8.1. Introduction

The continuously increasing speed and memory capacity of
supercomputers has, over the past decades, allowed for the creation of
ever more complex and accurate mathematical simulations. There are
challenges facing high performance computing (HPC) however. Chief
among these are the monetary and environmental costs involved in

purchasing and running a HPC system. The electricity costs alone for an exascale supercomputer are estimated to be more than $80 million [Ref. 1] a year.

To create a supercomputer that achieves the maximum possible performance for a given power/space budget, the architecture of the system needs to be tailored to the applications of interest. This involves optimally balancing resources such as memory, data storage and networking infrastructure based on detailed analysis of the applications. As well as these high-level optimizations, the architecture of the chips in the system needs to provide both speed and low power consumption.

Currently the top 500 supercomputers [Ref. 2] are built from relatively general purpose servers which rely on CPUs (and more recently general purpose GPUs) for computation. The architectures used by these chips are suitable for a wide range of tasks however this also means that their low-level architectures are not necessarily optimal for the applications the supercomputer is designed to run. Figure 8.1 shows how little of a modern CPU is dedicated to actual

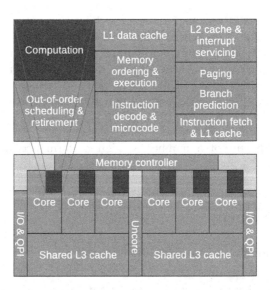

Fig. 8.1.   Simplified diagram of an Intel Westmere 6-core processor chip, highlighting the approximate portion of the chip performing computation versus other functions.

computation. The rest of the chip is dedicated to subsystems such as caches, branch predictors and schedulers designed to speed up programs. Far higher performance and efficiency can be had by designing the architecture such that it is a perfect fit for an application.

The underlying architecture of a computer system can be optimized by developing application-specific integrated circuits (ASICs). Designing ASICs is, however, a very costly exercise and limits how the supercomputer can be adapted and improved over time. Peter Cheung has made many contributions to reconfigurable computers based on chips such as field programmable gate arrays (FPGAs); they are a lower cost way of unlocking the gains that architectural customization can bring while retaining the programmability that makes general purpose computing chips so popular.

8.2. Dataflow Technology

Maxeler's dataflow computing paradigm is fundamentally different to computing with conventional CPUs. The technology is an evolution of dataflow computer [Ref. 3] and systolic array processor [Ref. 4] concepts developed in the 1970s and 1980s.

Figure 8.2 shows how computing with CPU cores compares to the model used in a dataflow computer. Rather than constantly fetching and writing to the main memory, data is read once and then moved through dataflow cores placed exactly where required. The underlying architecture is adapted to the application and the data movement is minimized.

8.2.1. *Dataflow engines*

Maxeler has developed *dataflow engines* (DFEs) to provide a high-performance reconfigurable dataflow computing platform. A DFE has at its heart a modern high-speed reconfigurable chip. Wired to this chip is a significant quantity of memory for storing application data and various high bandwidth interconnects that allow the DFE to communicate quickly with other DFEs, networks and general purpose computers.

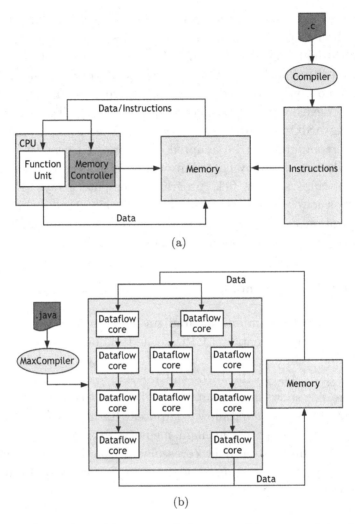

Fig. 8.2. Computing with a control flow core (a) compared to dataflow cores (b).

The reconfigurable chips used so far in DFEs have been FPGAs. FPGAs have a very flexible fabric in which dataflow graphs can be emulated. The development of reconfigurable integrated circuits developed specifically for DFEs could bring significant performance improvements to applications and improve programmability. There are also modifications that could be incorporated into FPGAs to make them more suitable for dataflow applications. The buffers used to enforce schedules in dataflow graphs, for example, are currently implemented using dedicated FPGA memory resources such as block RAMs with separate addressing logic implemented in configurable logic blocks (CLBs). This creates routing congestion around the memory block and reduces the frequency at which they can be clocked. Autonomous memory blocks [Ref. 5] could solve this problem by creating a dedicated configurable addressing circuit within the memory block itself. As well as reducing congestion this would reduce the amount of work required from compilation tools to place the CLBs.

8.2.2. *MaxCompiler*

DFE configurations are designed in the Java-like *MaxJ* language and are compiled using MaxCompiler. Figure 8.3 shows how MaxCompiler fits into the compilation system used for an application.

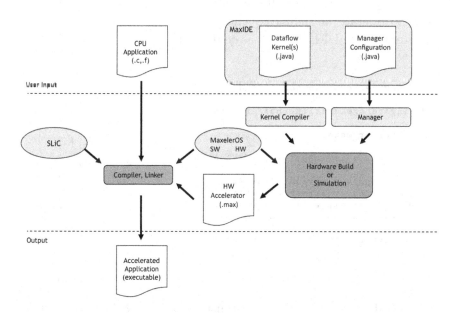

Fig. 8.3.   Compilation flow of a C/Fortran based MaxCompiler application.

MaxCompiler designs use an architecture which is conceptually similar to the globally asynchronous locally synchronous (GALS) architecture previously used in FPGA design [Ref. 6]. Designs consist of a number of locally synchronous *kernels* connected together and to other asynchronous DFE resources using the *manager*. This architecture allows the clock rate for each kernel to be individually tweaked to provide the optimum data throughput vs. routability tradeoff.

Once a configuration has been compiled and loaded onto a DFE it is communicated with using a simple automatically generated API accessible from a variety of languages such as C, R and Python.

8.2.2.1. *Kernels*

Kernels are at the heart of the MPC concept. Kernels describe computation and data manipulation in the form of dataflow graphs.

MaxCompiler transforms kernels into fully pipelined synchronous circuits. Figure 8.4 shows how a simple MaxJ kernel maps to a dataflow graph. The circuits are scheduled automatically to allow them to operate at high frequencies and make optimal use of the resources on the DFE. As well as enabling the maximum exploitation

```
class MovingAverageKernel extends Kernel
{
  MovingAverageKernel(KernelParameters parameters)
  {
    super(parameters);

    DFEVar x = io.input("x", dfeFloat(8, 24));

    DFEVar prev = stream.offset(x, -1);
    DFEVar next = stream.offset(x, +1);
    DFEVar sum = prev + x + next;
    DFEVar result = sum / 3;

    io.output("y", result, dfeFloat(8, 24));
  }
}
```

(a)   (b)

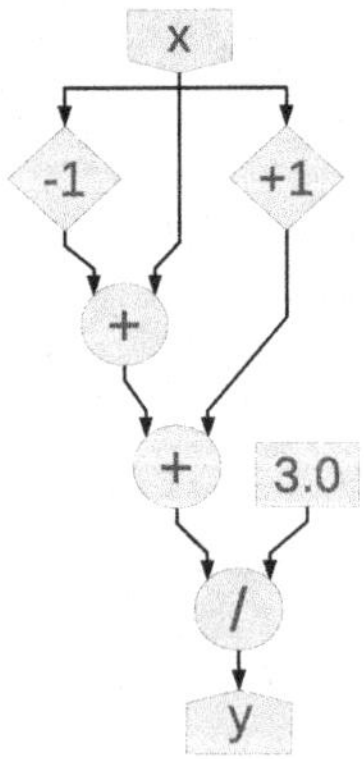

Fig. 8.4.   MaxJ source code for (a) a simple moving average kernel and (b) the resulting generated dataflow graph.

of low-level parallelization the fully pipelined nature of kernels allows them to be easily modeled at a high level which helps to avoid the wasted programming effort sometimes seen with other less predictable programmable systems.

8.2.2.2. *The manager*

The manager describes the asynchronous parts of a DFE configuration which are connected together using a simple point-to-point interconnection scheme [Ref. 7]. The manager allows developers to connect resources such as memory and PCIe streams to kernels to keep them fed with data.

8.2.2.3. *SLiC (Simple live CPU) API*

DFEs are controlled from a CPU using the SLiC API. This API comes in a variety of flavours ranging from the basic and advanced automatically generated static APIs to the more flexible dynamic API. As well as a C API MaxCompiler can generate bindings in a variety of other languages so that DFEs can be used to accelerate portions of Matlab, R and Python code without the need for users to write their own wrappers.

8.3. Financial Valuation and Risk Analysis

There is a need in the finance industry to compute ever more complex mathematical models in order to accurately price instruments and calculate exposure to risk. These applications often rely on complex control flow which can be difficult to accelerate using high performance single-instruction multiple-data (SIMD) architectures.

8.3.1. *Tranched credit derivatives*

Credit derivates are financial instruments used to transfer some of the risk associated with an asset, such as a bond, to an entity other than the lender in return for regular payments.

Collateralized default obligations (CDOs) are a type of credit derivative created from a pool of assets. This portfolio is divided into layered tranches. Each tranche represents a risk class; *junior* tranches have a high risk of default and so provide higher returns whereas *senior* tranches have a low risk of default and provide lower returns. This differential exists because if the value of the underlying assets drop, investors holding the *senior* tranches will be paid before those holding *junior* tranches.

Dataflow technology has been used to accelerate CDO valuation and shows a 31 × speedup when compared to 8-core Xeon servers [Ref. 8]. The *random-factor loading model* [Ref. 9] is used to value the CDO tranches. This application highlights the power savings that can be acheived using dataflow technology, acheiving the 31 × speedup while using only 6% more power than the reference system.

8.3.2. *Interest rate derivatives*

Interest rate derivatives are financial products which allow investors to incorporate interest-rate-based instruments into their portfolios. One example of an instrument which can be the underlying asset behind an interest rate derivative is an *interest rate swap*. In a swap one party is commonly paying a floating rate of interest linked to the interest rate set by an authority such as a central bank and a second party pays a fixed rate of interest. Often the fixed rate interest is paid directly to the party paying the floating rate. In this case the first party is taking a risk in return for the opportunity to make a profit should the floating rate be lower on average than the fixed rate.

Calculating the value of derivatives based on instruments such as swaps and other assets is complicated and needs to take into account a wide range of variables such as foreign exchange, interest and equity rates. It is impractical for financial institutions to simulate every possible scenario and so the Monte Carlo method is often adopted, simulating many possible scenarios and combining the results to calculate a price.

The Monte Carlo method has been used as the basis for a derivative pricing system which achieved wall clock speedups of up to 200 × versus single cores [Ref. 10]. The use of Maxeler's dataflow technology meant that a complex design could be developed incorporating:

- Multi-asset classes.
- Flexibility to accommodate idiosyncratic payoff functions.
- Accurate and stable finite difference risk calculations.

8.4. Geophysics

Dataflow technology is well suited to a number of geophysics applications and speedups of approximately 200 × compared to single CPU core implementations have been recorded [Ref. 11, 12]. Geophysics applications often involve manipulating large datasets and many of the algorithms used can be concisely described in kernels. Applications based on fast Fourier transforms [Ref. 13] and explicit finite difference [Ref. 14] have been successfully implemented using dataflow technology.

8.4.1. *Reverse time migration*

Reverse time migration (RTM) is a computationally intensive algorithm used in the oil and gas industry for subsurface imaging.

As part of the exploration process acoustic waves are broadcast through the Earth's surface and the reflections recorded. Geophysicists use these reflections to create models of structures beneath the ground. Computers running RTM are then used to model the wave propagation forwards from the source (mirroring the physical process) followed by the propagation of the reflections backwards from the receiver (the reverse of the physical process). The two are correlated and combined into an image which is then used to further refine the earth model.

Finite difference is a well known way of solving the partial differential equations used to model acoustic wave propagation. Figure 8.5 shows a basic form of the algorithm used. RTM based on finite difference and using Maxeler's MaxGenFD tool has been accelerated using dataflow technology, achieving speedups of around

```
curr = zeros()
prev = zeros()
for t = 0 to tmax:
  for i = 0 to X*Y*Z-1:
    l = convolve(curr[i], stencil)
    next[i] = 2 * curr[i] - prev[i]
                  + vv[i] * l
    next[i] += stimulus[t, i]
    apply_boundary_condition(curr, next)
    swap(prev, curr, next)
```

Fig. 8.5.   Pseudo-code for a finite difference wave propagator.

200 × when compared to single CPU cores [Ref. 15]. These speedups enable the use of higher frequency modeling improving the quality of the results.

8.4.2. *MaxGenFD*

MaxGenFD [Ref. 14] is a framework built on top of MaxCompiler which simplifies the development of high performance explicit finite difference solvers for seismic processing applications. MaxGenFD manages domain decomposition, boundary conditions, convolution and data set handling. Applications such as forward modeling, RTM and full waveform inversion can be quickly developed and optimized using MaxGenFD and put into production.

Figure 8.6 shows how data is moved through the DFE in a typical MaxGenFD application. The Earth model remains constant throughout a simulation whereas the wavefields are updated as part of each timestep.

MaxGenFD extends the flexible stream typing system provided by MaxCompiler (first showcased by Maxeler as part of the Photon compiler [Ref. 16]). The typing system allows developers to prototype a design using floating point types and then quickly transition to a variable-width fixed point implementation of the same design. Additionally MaxGenFD globally scales fixed point values at run-time to maximise precision. One of the limitations of this technique is that geophysicists need to know the relative maximums of the

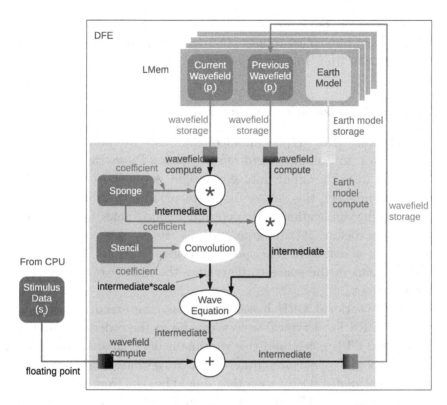

Fig. 8.6.   Diagram showing the flow of data in a typical MaxGenFD application.

wavefield inputs to their kernel in order to maximise their individual precision. Lightweight dynamic scaling systems such as *Dual FiXed-point* [Ref. 17] could be adopted in the future to make this system more flexible without resorting to computationally expensive floating point operations. For example the second exponent could be used to catch peaks in the wavefield allowing the bulk of the wavefield to be computed in a higher precision using the first exponent.

Research into multiple-wordlength optimization (for example [Ref. 18]) is ongoing and is important for tools such as MaxGenFD as it allows them to make the best of use of the available resources on reconfigurable chips. As MaxGenFD designs are a relatively high-level description of a mathematical algorithm it might even be possible in the future to automatically work out the optimal types given an acceptable error constraint. A similar concept been demonstrated for floating point implementations of Fourier transforms using *automatic differentiation* [Ref. 19].

8.4.3. *CRS seismic trace stacking*

CRS stacking [Ref. 11] is an algorithm used to process seismic survey data to compute *zero-offset* traces. These can be easily computed from the input data given eight parameters to the CRS equation, however the values for these parameters are unknown. The stacking application must determine good values for the eight parameters before it can compute the stack. This is done by performing a search in 8-D space. A fitness function is evaluated at each point in the space to determine the quality of the current parameter set.

On conventional CPU based computers, the execution time to compute CRS for a typical survey can be of the order of 1 month using 1,000 CPU cores, and this runtime is dominated by the computation of the *semblance* (fitness) and traveltime. The traveltime function calculates the location that should be read from a data trace based on the eight CRS parameters, while the semblance function computes on a window of data values from that location.

The application is accelerated by moving the semblance and traveltime functions onto a DFE. The code which controls the search through the parameter space remains untouched on the CPU.

The extra computational capability of the DFEs compared to the CPU means that some algorithmic changes are required to reach the full potential performance. When a single search is performed many data items are read, but relatively little computation is performed on each item. By transforming the algorithm so multiple sets of computation are performed in parallel the computational intensity can be increased and this allows the DFEs performance to be fully exploited without being limited by memory bandwidth.

With both semblance and traveltimes computation on the DFE, over 99% of the total runtime from the CPU is accelerated. The final implementation gives a speedup of approximately 230 × compared to a single core for land datasets and 190 × for marine datasets [Ref. 11], approximately 30 × greater performance per watt.

8.5. Conclusions

Research has shown that dataflow computers using reconfigurable logic can be used to both significantly improve the speed at which computation can be performed while also reducing power consumption when compared to traditional computer technology. Dataflow technology is a proven way to push the performance of reconfigurable computers while maintaining an easy to use and flexible development system.

References

1. O. Pell and O. Mencer. Surviving the End of Frequency Scaling with Reconfigurable Dataflow Computing, *SIGARCH Computer Architecture News*, 39(4), 60–65, 2011.
2. Top 500 Supercomputer Types. [Online] Available at: http://www.top500. org. [Accessed 4 April 2013].
3. J.B. Dennis. Data Flow Supercomputers, *Computer*, 13(11), 48–56, 1980.
4. H.T. Kung. Why Systolic Architectures?, *Computer*, 15(1), 37–46, 1982.
5. W.J.C. Melis, P.Y.K. Cheung, and W. Luk. Autonomous Memory Blocks for Reconfigurable Computing, in *Proc. International Conference on Circuits and Systems*, vol. 2, pp. 581–584, 2004
6. A. Royal and P.Y.K. Cheung. Globally Asynchronous Locally Synchronous FPGA Architectures, in *Proc. International Conference on Field Programmable Logic and Applications*, pp. 355–364, 2003.
7. T.S.T. Mak *et al*. On-FPGA Communication Architectures and Design Factors, in *Proc. International Conference on Field Programmable Logic and Applications*, pp. 1–8, 2006.
8. S. Weston *et al*. Accelerating the Computation of Portfolios of Tranched Credit Derivatives, in *Proc. Workshop on High Performance Computational Finance*, pp. 1–8, 2010.
9. L. Andersen and J. Sidenius. Extensions to the Gaussian Copula: Random Recovery and Random Factor Loadings, *Journal of Credit Risk*, 1(1), 29–70, 2004.
10. S. Weston *et al*. Rapid Computation of Value and Risk for Derivatives Portfolios, *Concurrency and Computation: Practice & Experience*, 24(8), 880–894, 2012.
11. P. Marchetti *et al*. Fast 3D ZO CRS Stack An FPGA Implementation of an Optimization Based on the Simultaneous Estimate of Eight

Parameters, in *Proc. European Association of Geoscientists and Engineers Conference,* 2010.

12. T. Nemeth *et al.* An Implementation of the Acoustic Wave Equation on FPGAs, *SEG Technical Program Expanded Abstracts,* 27(1), 2874–2878, 2008.

13. C. Tomas *et al.* Acceleration of the Anisotropic PSPI Imaging Algorithm with Dataflow Engines in *Proc. Annual Meeting and International Exposition of the Society of Exploration Geophysics,* 2012.

14. O. Pell *et al.* Finite Difference Wave Propagation Modeling on Special Purpose Dataflow Machines, *IEEE Transactions on Parallel and Distributed Systems,* 24(5), 906–915, 2013.

15. D. Oriato *et al.* FD Modeling Beyond 70Hz with FPGA Acceleration, *in Proc. SEG HPC Workshop,* 2010.

16. J.A. Bower *et al.* A Java-based System for FPGA Programming, in *Proc. FPGA World Conference,* 2008.

17. C.T. Ewe, P.Y.K. Cheung, and G.A. Constantinides. Dual Fixed-point: an Efficient Alternative to Floating-point Computation, in *Proc. International Conference on Field Programmable Logic and Applications,* pp. 200–208, 2004.

18. G. Constantinides, P.Y.K. Cheung, and W. Luk. The Multiple Wordlength Paradigm, in *Proc. International Symposium on Field-Programmable Custom Computing Machines,* pp. 51–60, 2001.

19. A. Gaffar *et al.* Floating-point Bitwidth Analysis via Automatic Differentiation, in *Proc. International Conference on Field-Programmable Technology,* pp. 158–165, 2002.

Chapter 9

Future DREAMS: Dynamically Reconfigurable Extensible Architectures for Manycore Systems

Lesley Shannon

School of Engineering Science,
Simon Fraser University

In today's world of cloud computing and petascale computing facilities (with discussions of exascale computing system design), manycore systems are indeed a reality. The challenge of increasing computational power, while managing/reducing power consumption, is leading to the consideration of new compute solutions. This chapter proposes how reconfigurability may become part of the solution for these new systems and whether it is time to reconsider some of the underlying components in traditional computing system design.

9.1. Introduction

Since the sale of the first commercially available microprocessors in 1971 (Intel's 4004) [Ref. 1], computing system design has been on an exponential trajectory, leveraging the exponential increase in transistor counts and clock frequencies. Innovations range from RISC instruction sets, to caches, to branch predictors; however, the fundamental CPU architecture proposed by Von Neumann and others has changed very little. Although more innovative architectures have been proposed (e.g. dataflow machines, dynamically reconfigurable computers), they have not garnered the commercial success of those based on the original Von Neumann style architecture.

Recently, computing design innovations have come in the form of multicore/manycore architectures and networks-on-chip, stamping out an increasing number of traditional computing cores on a single die to increase performance. However, this has not lead to corresponding increases in computational efficiency due to the overheads and challenges inherent in parallel programming and resource sharing. Given the future of systems with thousands/hundreds of thousands of processing elements, we have an opportunity to re-examine computing system design and perhaps revisit some of the architectural fundamentals that have changed minimally over the past forty years.

This chapter questions the future of computing system design for manycore systems and discusses the place of reconfigurability and reconfigurable computing in this future. First, is a brief description of computing innovation in the 1900s. Next, the changes and challenges of the past twelve years are presented. Finally, a vision for the future of computing and the opportunities and problems that lay ahead are outlined.

9.2. Yesterday

Since the first commercial CPUs of the 1970s, computer architectures have been on an exponential growth track, leveraging Moore's Law. Datapaths grew from 4 bits to 8 bits, followed by 16 bits, 32 bits, and now 64 bits (even 128 in specialized cases). Additional highlights from these many years of research as to how to improve the efficiency and processing throughput of a single processor's architecture include investigations of:

- effective instruction set architectures (ISAs) (e.g. CISC versus RISC),
- pipelining,
- how to leverage additional on-chip memories (i.e. caches),
- how to increase a system's memory using virtual memory, and
- how to efficiently guess what to do when encountering a conditional branch (i.e. branch predictors).

The impact of shrinking process technology sizes, resulting in the exponential increase in the number of transistors that can be fabricated on a chip with each new generation, combined with all the other new technological advances has had an astounding impact on computing system design. As a simple example, the ever popular Commodore 64 [Ref. 2] from the early 1980s had 64 kB of memory, whereas today's personal computing systems easily have a terabyte worth of system memory (increasing by more than 15 million times in size over the last 30 years) and 8 GB of main memory (an increase of more than 130 thousand times). Computing clock frequencies have also similarly increased, if at a somewhat slower rate, up until 2004. Concurrently, improvements in the accompanying software technology (e.g. compilers and operating systems) have provided abstractions that have made these platforms accessible to a large user base.

9.2.1. *Alternative computing system designs*

In conjunction with the more traditional avenues of investigation, researchers have also explored other computing models that relied more on spatial computing than on the temporal sharing of resources. One form of spatial computing examined *dataflow* architectures such as data process networks [Ref. 3], Kahn process networks [Ref. 4], and YAPI [Ref. 5]. These architectures, like systolic arrays, assume that computing elements are networked together via communication links that are used to move data through the system. Although this style of architecture is very useful for data-intensive applications (e.g. streaming applications), they are not good for applications that have numerous conditional branches as the resources for both options in the branch must be allocated even though they are never used. Furthermore, unless the connectivity pattern between computing elements can be dynamically reconfigured between applications, the architecture is effectively hard coded to perform one set of operations.

Inspired by the reconfigurable hardware platforms provided by field programmable gate arrays (FPGAs), researchers have also considered dynamically reconfigurable computing architectures. These architectures have ranged from dynamic arrays of computing elements that can be reconfigured at a relatively fine grain for bit-wise operations [Refs. 6 and 7], to more coarse grain fabrics reconfigurable fabrics that operate on data words [Refs. 8 and 9]. These reconfigurable fabrics can then be integrated into the computing system as a reconfigurable functional unit (RFU). The RFU may be tightly integrated into the processor's pipeline [Refs. 6 and7] or more loosely integrated as a co-processing unit that can operate somewhat independently of the rest of the CPU [Refs. 8 and 9].

Since the spatial portion of these computing architectures is dynamically reconfigurable, their resources can be adapted and reused to solve any compute problem, making them more area efficient than a fixed dataflow architecture. However, reconfiguration comes at a cost in terms of time and space. The time required to reconfigure a fabric is directly proportional to the granularity of its configuration. Fine-grain fabrics require the largest number of configuration bits and, therefore, have the largest area overhead for their configuration infrastructure as well as the longest configuration times. Conversely, coarse-grain fabrics need fewer configuration bits and do not require as long to reconfigure. However, their area overhead costs arise from unusable resources or inefficient use of the fabric as the additional configuration bits in fine-grain fabrics make it easier to customize them more precisely to the needs of the computation.

Although reconfigurable computing architectures can accelerate certain applications, the overhead of dynamic reconfiguration is generally too costly to make this approach worthwhile. As such, in spite of the potential for hardware acceleration, dynamic partial reconfigurable computing architectures have not had much success to date.

9.2.2. *Summary*

From one perspective, even a traditional CPU can be viewed as dynamically and partially reconfigurable: the instruction decoder "reconfigures" the CPU's datapath for each instruction to be executed. The key is that this configuration is so coarse-grain that the "configuration bits" (i.e. machine code instruction) can be stored in relatively few bytes and the datapath can be "reconfigured" on a cycle by cycle basis for each instruction. Single-instruction-multiple-data (SIMD), multiple-instruction-multiple-data (MIMD), and instruction windows enable the CPU to share its datapath for the execution of multiple instructions concurrently, to increase instruction level parallelism. Furthermore, their programming models and compiler flows are considerably simpler than those of actual dynamic partially reconfigurable computing architectures as their parallelism is limited. However, this restricted parallelism, due to limited computing resources and lack of application-specific customization, means they cannot compare with the efficiency of application execution using custom hardware.

9.3. Today

Today's high-performance processors rely on parallelism in lieu of increasing clock rates to improve computing efficiency. Not only can threads share a CPU's pipeline through symmetric multi-threading, but most modern processors have multicore architectures, where multiple CPU cores (two, four, or eight cores) are stamped out on the same die. Although these architectures originally had a shared bus format, network-on-chip architectures are becoming more prevalent with the increasing number of cores (16, 32, 64).

9.3.1. *Manycore computing*

As these systems become larger (hundreds/thousands of cores) and more prevalent, sharing resources efficiently among users, applications, and threads becomes more challenging. Realizing that the cost of moving data around the network from different "domains of coherency" can more than double execution time [Ref. 10], not to mention energy consumption, means that efficient methods of dynamically allocating resources to different applications are essential. To make these complex platforms more accessible to users, abstractions such as the "cloud" have been created. However, recent reports indicate that information technology represents 10% of the world's electricity consumption [Ref. 11]. Therefore, as we look forward, we need to question how we design these systems to be both computationally and *energy* efficient.

Based on power consumption alone, exascale computing systems will not be able to be built from solely high performance processors. One possibility is to build these new systems using low end processors with lower power consumption, such as those popular in embedded computing devices. Another possibility, is to consider a more heterogeneous system, with FPGAs to implement hardware accelerators, general purpose graphic processing units (GPGPUs) for floating-point-intensive modules, and general purpose processors. Although programming and allocating resources in a homogeneous system is an easier problem, the heterogeneous solution may provide better overall performance at comparable costs, assuming the appropriate software infrastructure is in place.

9.3.2. *Challenges and opportunities*

Manycore systems provide both challenges and opportunities for researchers and programmers. What types of computing components should they comprise? What type of synthesis and compilation flow is needed to map designs to these systems? How will the system's scheduler adapt to the various workloads? What additional software infrastructure/changes in programming models are needed to expose an application's parallelism and facilitate its mapping to a manycore system's resources. The remainder of this chapter will focus on discussing a vision of a solution to some of these challenges.

9.4. A Vision of Tomorrow

This section proposes a vision for the programming, compilation, scheduling, and design of a dynamically reconfigurable extensible architecture for manycore systems (DREAMS). To allow for the greatest flexibility in this model, it is assumed that these systems may be comprised of heterogeneous computing (processing) elements.

9.4.1. *Programming and compilation*

The programming model for the proposed system should rely on high-level languages. The model should support domain-specific languages as well as the more generic ones popular with programming experts and novices. Although not all languages may be compiled down to the same efficiency of execution on the platform, they should all work. Instead, the expectation is that great advances in compiler technology are needed.

Specifically, the compiler will have to uncover parallelism and generate executable subcomponents that can be executed on multiple processing element (PE) architectures to support the platform's heterogeneity. As shown in Fig. 9.1(a), current synthesis/compilation flows translate multiple source modules into a single executable format for a specified platform. However, as shown in Fig. 9.1(b), the proposed solution for heterogeneous manycore systems is to compile the source modules into a set of executable modules. In this scenario,

(a) Current many-to-one Compilation/Synthesis Flow Model
(many source files to one executable)

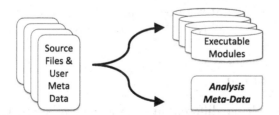

(b) Proposed many-to-many Compilation/Synthesis Flow Model
(many source files to multiple executable modules and static analysis meta-data)

Fig. 9.1. (a) Current synthesis and compilation model. (b) Proposed synthesis and compilation model.

not only should these executable modules be mappable to a variety of PE types, but they should also be mappable to platforms with a varied number of PEs so that the design's final executable format remains portable to different platforms. Finally, it is expected that the compiler will accept additional input, along with the traditional program source files, that will be evaluated by an additional analysis phase to what is found in current compilers.

The additional input to the compiler will be in the form of metadata from the user. It can be used to indicate information ranging from the criticality of high speed execution to the number of PE resources a user wishes to allocate to the problem (as this may impact the cost of performing cloud-based processing). Additionally, information such as compilation time and effort may be included along with the user's perspective as to key or bottleneck processing subcomponents.

The additional analysis phase will be used to perform a static analysis of the program and the perspective libraries that it can use. Ideally, this will generate additional meta-data (i.e. the *analysis meta-data* shown in Fig. 9.1(b)) that can be used by the platform's scheduler to effectively allocate resources so as to increase processing efficiency. This information may affect the number and/or type of resources allocated to the application's execution. Finally, instead of only generating application threads based on the programmer's software description of the implementation, the compiler will be able to generate executable modules comprised of task-level threads of execution to be mapped to various PEs, adding the necessary synchronization primitives needed to to ensure that the final implementation is functionally equivalent to the original description.

9.4.2. *Scheduling*

To schedule multiple different applications on a heterogeneous fabric concurrently and efficiently, the scheduling software needs information about the programs currently being scheduled. As described in the previous section, this can come in the form of metadata about the program from the compiler and user. In fact, it is expected that this type of information will be embedded in the executable as a header that can be used to allocate PEs and schedule task execution.

To assist the scheduler and enable it to adapt unknown workloads at run time, run-time performance-profiling hardware should be built into the computing fabric. This can be used to obtain additional run time statistics about the application's threads — what resources they need, how efficiently they run on different types of PEs, etc. Ideally, these profiling resources can be used by both the programmer to understand the program's run time behaviour so as to better use processing resources and by the OS to adapt its resource allocation to the needs of all threads in the current workload.

9.4.3. *Proposed architecture*

In manycore systems with hundreds/thousands of processors, the cloud abstracts the "sea" of computing resources from the user. However, a hardware abstraction is also needed to at the execution level to "hide" the platform's complexity from the individual applications executing on it. Figure 9.2 provides a simple overview of the proposed DREAMS architecture with a homogeneous sea of PEs,

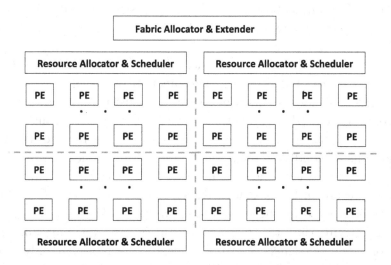

Fig. 9.2. A potential hierarchical architecture for a homogeneous sea of processing elements.

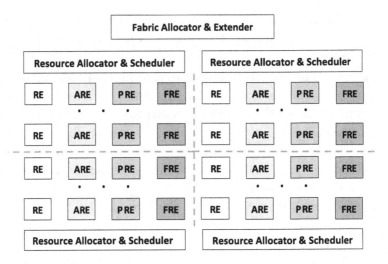

Fig. 9.3. A potential hierarchical architecture for a *heterogeneous* sea of processing elements.

reflective of current manycore systems. Ideally, this sea will instead be a heterogeneous mix of PEs, as shown in Fig. 9.3, comprising PEs with asymmetric configurations (APEs), or GPGPUs (GPEs), and even FPGAs (FPEs). This variety of processing elements ensures that different application types (e.g. streaming, floating-point intensive, etc.) can execute more efficiently by mapping them to their appropriate PE types.

The proposed architecture is hierarchical, with a *fabric allocator & extender* at the highest level. The function of this unit is twofold. First, it provides a method of connecting existing computing fabric resources to newly added units to *extend* the computing fabric, without requiring all of the PEs themselves to be aware of the new additions. Secondly, it reads the metadata in the header, *and only the header*, and uses the profiling data in the header in conjunction with a record of the workload currently allocated to the fabric and being monitored by the *resource allocators & schedulers*.

To ensure that the *fabric allocator & extender* unit does not become the bottleneck for processing on the fabric, it only reads the data in the application executable's header; as this is a point of serialization, it does not read any of the executable code, but directly

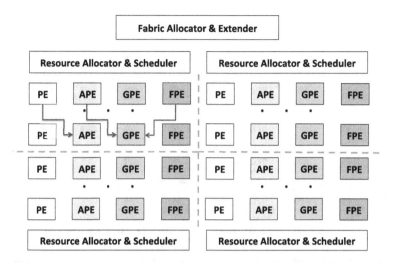

Fig. 9.4. An application-specific communication network within the PE fabric dynamically generated by the resource allocator.

routes it to the second tier of the hierarchy, the *resource allocator and scheduler* modules. The resource allocator and scheduler uses a portion of the application header to determine how many resources of each type it would like to allocate to the application and to generate an application-specific communication network between the PEs (see Fig. 9.4). If a resource allocator would like additional resources to allocate to the workload it is managing, it can request that the fabric allocator either: (1) reassign some of the workload to another portion of the fabric managed by a different resource allocator, or (2) get a resource allocator to share some of its PE resources with an application running on the adjacent PE fabric controlled by an adjacent resource allocator (see Fig. 9.5).

The resource allocator & scheduler uses the information in the application header in conjunction with the information on the workload it is currently managing on its portion of the fabric to decide how many and what types of PEs to allocate to the application. Depending on the nature of the workload it is executing, the PE resources may be shared between threads and workloads. For each application, the resource allocator & scheduler will use the on-chip performance-profiling hardware to monitor the workload

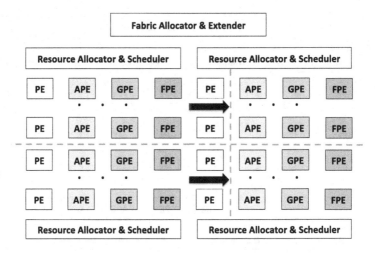

Fig. 9.5.   Resource allocators extending their fabric for application execution by "borrowing" PEs from an adjacent resource allocator.

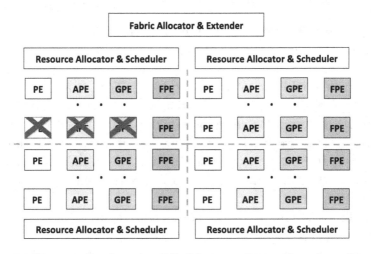

Fig. 9.6.   Dynamically detecting PE failures and removing them from the resource allocator's list of usable processing resources.

executing on the PEs in its fabric. This data will be recorded on a per-application basis and to enable the allocator to adapt to its workload's needs at runtime. Furthermore, this data will be aggregated at a workload level and fed back to the fabric allocator & extender to help the upper level of the platform ensure a balanced workload across the fabric. Finally, the profiling hardware will be used to detect failures in PEs. The resource allocator will mark these devices as offline (see Fig. 9.6) and ensure that no future applications are scheduled to use them as processing resources.

The final element of the processing infrastructure are the PEs. The PEs in the fabric are responsible for executing the tasks for the workload allocated to the resource allocator. The PEs may be heterogeneous or homogeneous with respect to each resource allocator. However, to ensure better locality and faster communication and data throughput, the tasks will be allocated spatially to the PEs and the underlying communication fabric that connects the PEs will be configured specifically for each application (recall Fig. 9.4). This fabric can be reconfigured on a per application basis and the network and PE resources can be time-shared among applications.

9.5. Conclusions

There are numerous research opportunities that become apparent when trying to envision how to make a sea of processing units work effectively and efficiently. The abstraction of the cloud provides users with a way to use these resources that hides the hardware. However, the question of what that hardware should really look like is an important consideration as is the design flow required to facilitate their programming and ensure their efficient operation.

While repeatedly stamping out the same resources or simply combining hundreds or thousands of the same chips into a fixed structure (i.e. a homogeneous solution) might be easy to build, it is not efficient (in terms of power and performance) as it cannot adapt. Furthermore, given the cost of building these systems, it is important to ensure that they are able to cope with component failures, etc. and to ensure their longevity and reliability for their users. These types of computing systems will not only have to adapt to their applications but they will also have to be able to self-monitor and self-heal (or at the very least adapt to unit failures so that the system operates efficiently until it is repaired). Addressing these research challenges promises to provide another 40 years of exciting innovation in computing system design.

Acknowledgements

Peter: Your drive, enthusiasm, body of work, and life balance are inspirational. You are a wonderful mentor, edifying even the most confused soul. Your generosity of time and spirit are uplifting and my gratitude for your friendship is immeasurable.

In short, "Thank you Peter, for everything." I am greatly indebted to you for hosting me at Imperial College London and giving me this opportunity to work with so many wonderful people; this time and experience are priceless. I know that my sabbatical here will be a memory that I will cherish forever.

I look forward to being your friend for many years to come and wish you many happy and exciting adventures in the future. All the best on this special day; I hope we, your friends, are able to convey at least some of the joy you bring into our lives. My best always, Lesley.

References

1. The Story of the Intel 4004. [Online] Available at: http://www.intel.com/content/www/us/en/history/museum-story-of-intel-4004.html. [Accessed 23 April 2014].

2. Commodore 64 — 1982. [Online] Available at: http://oldcomputers.net/c64.html. [Accessed 23 April 2014].

3. E. Lee and T. Parks. Dataflow Process Networks, *Proc. IEEE*, 83(5), 773–799, 1995.

4. G. Kahn. The Semantics of a Simple Language for Parallel Programming, in *Proc. IPIF Congress*, 1974.

5. E. Kock *et al.* YAPI: Application Modeling for Signal Processing Systems, in *Proc. Design Automation Conference,* pp. 402–405, 2000.

6. S. Hauck *et al.* The Chimaera Reconfigurable Functional Unit, *IEEE Transactions on VLSI Systems*, 12(2), 206–217, 2004.

7. T.J. Callahan, J.R. Hauser, and J. Wawrzynek. The Garp Architecture and C Compiler, Computer, 33(4), 62–69, 2000.

8. S. Goldstein *et al.* PipeRench: A Coprocessor for Streaming Multimedia Acceleration, in *Proc. International Symposium on Computer Architecture,* pp. 28–39, 1999.

9. C. Ebeling, D.C. Cronquist, and P. Franklin. RaPiD: Reconfigurable Pipelined Datapath, in *Proc. International Workshop on Field-Programmable Logic and Applications*, pp. 126–135, 1996.
10. S. Zhuravlev, S. Blagodurov, and A. Fedorova. Addressing Shared Resource Contention in Multicore Processors via Scheduling, in *Proc. Conference on Architectural Support for Programming Languages and Operating Systems*, pp. 129–142, 2010.
11. J. Clark. IT Now 10 Percent of World's Electricity Consumption, Report Finds. [Online] Available at: http://www.theregister.co.uk/2013/08/16/it_electricity_use_worse_than_you_thought/. [Accessed 23 April 2014].

Chapter 10

Compact Code Generation and Throughput Optimization for Coarse-Grained Reconfigurable Arrays

Jürgen Teich, Srinivas Boppu, Frank Hannig and Vahid Lari

Hardware/Software Co-Design,
Department of Computer Science,
University of Erlangen-Nuremberg, Germany

We consider the problem of compact code generation for coarse-grained reconfigurable architectures consisting of an array of tightly interconnected processing elements, each having multiple functional units. Switching from one context to another one is realized on a cycle basis by employing concepts similar as in programmable processors, i.e. a processing element possesses an instruction memory, an instruction decoder, and a program counter and thus can be seen as a simplified VLIW processor. Such architectures support both loop-level parallelism by having several processing elements arranged in an array and instruction-level parallelism by their VLIW nature. They are often limited in instruction memory size to reduce area and power consumption. Hence, generating compact code while achieving high throughput for such a processor array is challenging. Furthermore, software pipelining allows such an array to execute more than one iteration at a time, making the code generation and compaction even more ambitious. In this realm, we present a novel approach for the generation of compact and problem size independent code for such VLIW processor arrays on the basis of control flow graph notations. For selected benchmarks, we compare our proposed approach for code generation with the Trimaran compilation infrastructure. As a result, our approach may reduce the code size by up to 64% and may achieve up to 8.4 × higher throughput.

10.1. Introduction

In high-performance embedded computing systems, *coarse-grained reconfigurable arrays* (CGRAs) can be used to speed up parts of an application. As an example, they can serve as loop accelerators for many signal processing algorithms and thus boost the overall performance of the application. Examples of CGRAs include, for instance, PACT XPP [Ref. 1] or ADRES [Ref. 2], both of which are arrays that can switch between multiple contexts by run-time reconfiguration. Other commercial processor array examples that contain small VLIW processors are the picoArray architecture [Ref. 3], which consists of 430 16-bit 3-way VLIW processors, or Adapteva's Epiphany architecture with up to 1,024 dual issue processors [Ref. 4].

Our work starts with the specification of an arbitrarily nested loop program in a high-level language and mapping it onto a CGRA architecture, respecting a set of given resource and throughput constraints. Here, one of the main challenges is to generate the programs (codes) for each of the computation-centric, but instruction-memory limited VLIW processors automatically. The general perception of VLIW processors is that they may suffer from an overhead in code size, e.g., when not always employing all functional units. Therefore, numerous software pipelining techniques such as [Ref. 5] have been proposed, which allow a single processor to execute more than one iteration at a time and thus optimize the resource utilization and consequently also the throughput. As a basis of our work, we consider an extension of a class of CGRAs called *WPPA (weakly programmable processor array)* as proposed by Kissler and others [Ref. 6]. Here, instead of computing loop bound control in each processor and producing corresponding overheads in each processor, a *controller* is proposed that generates branch control signals that are issued once and propagated to the individual receiving processors in a pipelined manner. This type and sharing of CGRA loop control thereby provides a unique feature of *multi-dimensional zero-overhead loop processing*.

We use partitioning techniques [Refs. 7–9] to map a given loop program onto a given fixed size processor array to meet the given resource constraints such as number of processors, number of functional units in each VLIW processor, register and program memory restrictions, etc. [Ref. 10]. Once partitioning is done, the whole array is divided into a set of *processor classes*, where all processors that belong to the same class execute the same assembly program only with possibly a delayed control. In the following, we consider the problem of *code size compaction subject to a given schedule of loop iterations* assigned to a processor. For reasons of scalability, two levels of hierarchy (symmetry) are exploited: first, such a minimization must only be done once for each processor class as each processor belonging to a class will be configured with the same binary code to execute. The code generation approach is therefore independent in complexity of the number of processors of a CGRA. Moreover, the compaction method proposed is shown to generate programs of length independent of the number of iterations to be executed. For this purpose, a special *control flow graph* will be introduced whose nodes represent blocks of instruction sequences and the edges represent the control flow between the blocks.

In summary, our contribution is an approach for compact code generation for loop programs targeting large scale CGRAs consisting of an array of VLIW processor cores and providing the following unique features:

- Scalability by reduction of the code generation problem to individual *processor classes* only instead of each processor (CGRA size problem independence).
- Single-core code generation and compaction preserving a given mapping and schedule of loop iterations.
- Generated code is independent in length of (loop) problem size by extraction and exploitation of repetitive instruction sequences called *program blocks* in an intermediate control flow graph specification (loop specification problem size independence).
- Exploitation of *multi-dimensional zero loop overhead* branching.

The rest of the chapter is organized as follows: in Sec. 10.2, our considered class of CGRA architecture is briefly explained. In Sec. 10.3, few basic definitions, our mapping flow, and background of the problem of code generation are explained. In Sec. 10.4, our approach of scalable and compact code generation and an algorithm to generate the outlined control flow graph are described. Case studies are presented and discussed in Sec. 10.5. Related work is given in Sec. 10.6, and finally, Sec. 10.7 concludes the chapter.

10.2. Coarse-Grained Reconfigurable Arrays

We consider CGRAs consisting of an array of VLIW processors arranged in a 1-D or 2-D grid with local interconnections as proposed in [Ref. 6], see also Fig. 10.1 [Ref. 11]. The main application domain of such CGRAs are highly compute-intensive loop specifications. Such tightly coupled processor arrays may exploit both loop as well as instruction parallelism while providing often an order of magnitude higher area and power efficiency than standard processors. The architecture is based on a highly parameterizable template,

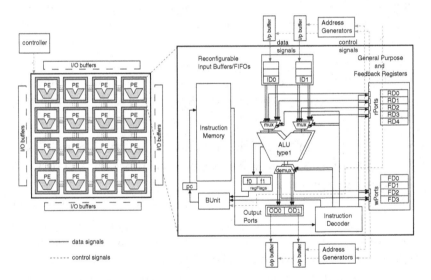

Fig. 10.1. Considered CGRA architecture template and inner structure of a customizable VLIW PE [Ref. 11], including multiple functional units, a VLIW instruction memory, and a set of registers for local data processing RD, input registers ID, output registers OD, and feedback registers FD for cyclic data reuse. The dark box surrounding each PE is called a *wrapper*. It allows the flexible configuration of the circuit-switched interconnections to neighbor PEs.

hence offering a high degree of flexibility, in which some of its parameters have to be defined at synthesis-time and some others can be reconfigured at run-time. For example, different types and numbers of functional units (e.g. adders, multipliers, logical units, shifters and data movers) can be instantiated as separate functional units. The processor elements (PEs) are able to operate on two types of signals, i.e. *data signals* whose width can be defined at synthesis-time and *control signals* which are normally one-bit signals used to control the flow of execution in the PEs. Therefore, two types of registers are realized inside the PEs: *data registers* and *control registers.*[a] These registers consist of four kinds of storage elements: general purpose registers, input and output registers, and feedback registers. The general purpose registers may store local data inside a PE and are named RDx in case of data and RCx in case of control registers. Input and output registers are mainly *ports* to communicate with neighboring PEs and are named IDx, ODx for data input/output registers, respectively, and ICx, OCx for control input/output registers. Input registers are implemented as shift registers of length l, whose *input delay* can be configured at run-time from 1 to l. So-called feedback data registers FDx are special rotating registers that are provided for the support of *loop-carried data dependencies* and *cyclic data reuse*. Each PE consists of a small *instruction memory* whose size can be defined at synthesis time. Moreover, a *multi-way branch unit* is employed in each PE, capable of evaluating multiple branch conditions, based on functional unit flags and control signals, in a single cycle. The branch unit takes one of 2^n targets in case of an n-way branch unit.

10.2.1. *Reconfigurable inter-processor network*

In order to provide support for many different interconnection topologies, a structure of multiplexers inside a so-called *wrapper* unit

---

[a]For the sake of better visibility, control registers and control I/O ports are not shown in Fig. 10.1. However, it should be noted that similar to data-path specific resources such as registers RDx, IDx, and so on, there exist also equivalent control path resources, e.g. registers RCx, ICx, etc.

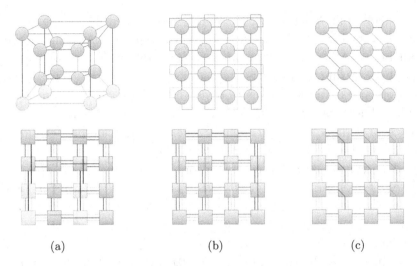

(a)   (b)   (c)

Fig. 10.2.   Different network topologies, such as, (a) 4-D hypercube, (b) 2-D torus, and (c) a systolic array, are implemented on a 4×4 processor array. In each case, the top figure depicts the logical connections between different nodes in the network, and the bottom figure illustrates the physical connections on the processor array established by configuration of interconnect wrappers.

[Ref. 6] around each PE is provided, which allows the reconfiguration of inter-PE connections flexibly. Thereby, many different network topologies, some of them shown in Fig. 10.2, may be realized. As illustrated in Fig. 10.1, the interconnect wrappers themselves are connected in a mesh topology.

Thanks to such a circuit-switched interconnect, a fast and reconfigurable communication infrastructure is established among PEs, allowing data produced in a PE to be used by a neighboring PE in the next cycle. To configure any particular interconnect topology, an adjacency matrix is specified for each interconnect wrapper in the array at synthesis-time. Each adjacency matrix defines how the input ports of its corresponding wrapper and the output ports of the encapsulated PE are connected to the wrapper output ports and the PE input ports. If multiple source ports are allowed to drive a single destination port, then a multiplexer with an appropriate number of input signals is generated [Ref. 6]. The select signals for such generated multiplexers are stored in configuration registers and can be changed even dynamically, i. e., different interconnect topologies — also irregular ones — can be established and changed at run-time.

Figure 10.2 shows some well-known network topology configurations such as 4-D hypercube, 2-D torus and connections of a systolic array implemented on a 4×4 processor array. In each case, the logical connection topology is depicted on the top and the corresponding array interconnect configuration is shown on the bottom of Figs. 10.2(a)–(c). Here, each interconnect wrapper has three connection channels to each neighboring PE. Depending on the desired topology, some of them are activated (shown in different colors) and some are inactive (shown in light gray color). It is worth mentioning that for some connections, wrappers may also be used to bypass a neighbor PE (connections with a length of more than one PE distance). This capability is highly useful, e.g. to implement a 2-D torus. As mentioned above, data may be transmitted on each port connecting two PEs within a single cycle. However, bypass capabilities are usually restricted to pass a maximum of two to four hops. Figure 10.2(c) shows how diagonal connections may be realized on CGRAs. This capability is used especially when optimally mapping (embedding) nested loop programs onto CGRAs which is the main focus of our work. In the next section, we therefore explain how loop applications may be systematically mapped and scheduled on CGRAs.

10.3. Mapping Applications onto CGRAs

10.3.1. *Front end*

In this section, we give an overview of the front end (see Fig. 10.3) of our approach for the mapping of loop nests onto massively parallel CGRA architectures. The approach strongly relies on the mathematical foundation of the polyhedron model [Ref. 12]. In Step (1), an algorithm is specified in *single assignment code* (SAC); hence the parallelism is explicitly given. The SAC is closely related to

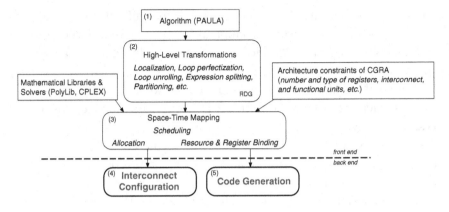

Fig. 10.3.   Flow for mapping a loop specification onto a CGRA.

set of recurrence equations called *piecewise linear algorithms* [Refs. 7 and 13] that are defined as follows:

Definition 10.1. (PLA) A *piecewise linear algorithm*, see, e.g. [Refs. 7 and 9] consists of a set of X quantified equations $S_1[I], \ldots,$ $S_X[I]$ where each $S_i[I]$ is of the form

$$x_i[P_i I + f_i] = \mathscr{F}_i(\ldots, x_j[Q_j I - d_{ji}], \ldots) \quad \text{if } \mathscr{C}_i(I) \; \forall \, I \in \mathscr{I}. \tag{10.1}$$

P_i, Q_j are constant rational indexing matrices and f_i, d_{ji} are constant rational vectors of corresponding dimension. $I \in \mathscr{I}$ is called *iteration vector* and $\mathscr{I} \subseteq \mathbb{Z}^n$ is an iteration space called *linearly bounded lattice* [Ref. 7] defining the set of iteration vectors I of the loop for which a quantified equation $S_i[I]$ is defined, respectively needs to be computed. Note that each quantified equation $S_i[I]$ may be defined over a different iteration space through the definition of *iteration-dependent conditionals* $\mathscr{C}_i$ in Eq. (10.1).[b] A PLA is finally called *piecewise regular algorithm* (PRA) if the matrices P_i and Q_j are the identity matrix. If in addition f_i is zero, an algorithm is in *output normal form*. In this case, the d_{ji} are also called *dependence vectors*,

---

[b]This is the reason why to call them piecewise linear, respectively regular.

which express a data dependency each from each variable instance $x_j[I - d_{ji}]$ to variable instance $x_i[I]$.

Finally, in order not to be restricted to loop programs with static data dependencies, *run-time dependent conditionals* have been introduced in [Refs. 10,14]. There, it was shown that a certain class of dynamic data dependencies may be expressed and incorporated into data-dependent functions of the form *ifrt(condition, branch1, branch2)*, extending PLAs to so-called *dynamic PLAs* or simply *DPLAs*.

For the textual representation of PLAs, we use the PAULA programming language [Refs. 10, 14] in our compiler for CGRAs. As our running example, we introduce an FIR filter example in the following.

Example 10.1. An FIR (finite impulse response) filter can be described by the simple difference equation $y(i) = \sum_{j=0}^{N-1} a(j) \cdot u(i-j)$ with $0 \le i < T$, N denoting the number of filter taps, T denoting the number of samples over time, $a(j)$ the filter coefficients, $u(i)$ the filter inputs, and $y(i)$ the filter results. The difference equation after parallelization, localization of data dependencies [Ref. 15], and embedding of all variables in a common iteration space can be written as the following PAULA program with the iteration domain $\mathscr{I} = \{I = (i\ j)^{\mathrm{T}} \in \mathbb{Z}^2 \mid 0 \le i \le T - 1 \ \wedge\ 0 \le j \le N - 1\}$.

```
program FIR
{ variable a_in 2 in integer<16>;
  variable u_in 2 in integer<16>;
  variable a 2 integer<16>;
  variable u 2 integer<16>;
  variable x 2 integer<16>;
  variable y 2 integer<16>;
  variable y_out 2 out integer<16>;
  parameter N;          // number of filter taps
  parameter T;          // number of samples
  par (i >= 0 and i <= T-1 and j >= 0 and j <= N-1)
  { S1: a[i,j] = a_in[i,j]        if (i==0);
    S2: a[i,j] = a[i-1,j]         if (i>0);
```

```
S3: u[i,j] = u_in[i,j]        if (j==0);
S4: u[i,j] = 0                if (i==0 and j>0);
S5: u[i,j] = u[i-1,j-1]       if (i>0 and j>0);
S6: x[i,j] = a[i,j]*u[i,j];
S7: y[i,j] = x[i,j]           if (j==0);
S8: y[i,j] = y[i,j-1]+x[i,j]  if (j>0);
S9: y_out[i,j] = y[i,j]       if (j==N-1);
   }
}
```

In Step (2) of our mapping flow depicted in Fig. 10.3, various high-level source-to-source transformations such as constant and variable propagation, common sub-expression elimination, loop perfectization, loop unrolling, expression splitting, dead-code elimination, affine transformations of the iteration space, strength reduction of operators (usage of shift and add instead of multiply or divide), and loop unrolling are performed. Algorithms with non-uniform data dependencies are usually not suitable for mapping onto regular processor arrays as they result in expensive global communication or memory. For that reason, a well-known transformation called *localization* [Ref. 15] exists, which allows the transformation of affine dependencies into uniform dependencies, i.e. a PLA into a PRA. In our mapping flow, a PRA is internally represented by a *reduced dependence graph* (RDG).

Definition 10.2. (Reduced dependence graph). For a given n-dimensional PRA, its RDG $G = (V, E, D, \mathscr{I})$ is a network where V is a set of nodes and $E \subseteq V \times V$ is a set of edges. Each edge $e = (v_j, v_i)$ corresponding to a data dependency is annotated with the corresponding dependence vector $d_{ji} \in \mathbb{Z}^n$. Nodes therefore represent variable names and there is an edge between v_j and v_i if and only if there exists a quantified equation $S_i[I]$ of the form $x_i[I] = \mathscr{F}_i(\ldots, x_j[I - d_{ji}], \ldots)$ if $\mathscr{C}_i(I)$.

Example 10.2. For the running FIR filter specification introduced in Example 10.1, the corresponding RDG is shown in Fig. 10.5(b). For each variable name, there is a node and for each data dependency, there is a corresponding directed edge. Apart from data dependence vectors (not shown in the figure), the RDG serves as an important intermediate data structure for subsequent code generation and will be decorated with much additional information in the course of compilation.

Next, we briefly explain *partitioning*, a high-level transformation that is used to map a given loop iteration space onto a fixed size processor array.

10.3.1.1. *Partitioning*

Loop partitioning is a well-known transformation which covers the iteration space of computation using congruent hyperplanes, hyper-quaders, or parallelepipeds called *tiles*. One purpose of partitioning is to map a loop nest of given size onto an array of processors of fixed size and therefore map tiles to processors and schedule these accordingly. However, the transformation has also been studied in detail for compilers with the purpose of acceleration through better cache reuse on sequential processors and known as *loop tiling or blocking* [Ref. 16]) targeting implementations on parallel architectures ranging from supercomputers to multi-DSPs and FPGAs. Well-known partitioning techniques are *multi-projection*, LSGP (local sequential global parallel, often also referred to as clustering or blocking) and LPGS (local parallel global sequential, also referred as tiling) [Ref. 8]. Formally, partitioning decomposes a given iteration space $\mathscr{I}$ using congruent tiles into the spaces $\mathscr{I}_1$ and $\mathscr{I}_2$, i.e., $\mathscr{I} \mapsto \mathscr{I}_1 \oplus \mathscr{I}_2$.[c] $\mathscr{I}_1 \subset \mathbb{Z}^n$ represents the points (iteration vectors) within the tile and $\mathscr{I}_2 \subset \mathbb{Z}^n$ accounts for the regular repetition of tiles, i.e. the space of non-empty tile origins.

Example 10.3. For the running FIR filter example, the iteration space is shown in Fig. 10.4 for an $N = 32$ tap filter. In order to map this filter onto a CGRA with four processors using LSGP

---

[c] For parallelotope-shaped tiles, $\mathscr{I}_1 \oplus \mathscr{I}_2 = \{I = I_1 + P \cdot I_2 \mid I_1 \in \mathscr{I}_1 \wedge I_2 \in \mathscr{I}_2 \wedge P \in \mathbb{Z}^{n \times n}\}$. Here, P is called tiling matrix.

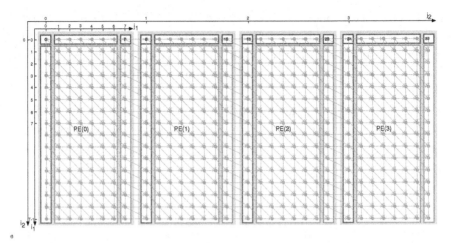

Fig. 10.4. Iterations of an $N = 32$ tap FIR filter including data dependencies partitioned to be executed on a linear 4 PE (1×4) CGRA using LSGP partitioning. For this purpose, the iteration space is divided into 4 tiles. The iterations of each tile will then be executed in a pipelined way by each PE. Shown also is a classification of iterations of the same type of instructions to be performed at each iteration by colors and named *program blocks*.

partitioning, the iteration space needs to be partitioned into exactly four tiles each being assigned to one processor. The tiling matrix chosen for partitioning the filter onto a 1×4 processor array is $P = \begin{pmatrix} T & 0 \\ 0 & 8 \end{pmatrix}$. For instance, it can be verified that the index point $I = (10\ 9)^{\mathrm{T}}$ is uniquely mapped to $I_1 = (10\ 1)^{\mathrm{T}}$, $I_2 = (0\ 1)^{\mathrm{T}}$, where $I_1 = (i_1\ j_1)^{\mathrm{T}} \in \mathscr{I}_1$ and $I_2 = (i_2\ j_2)^{\mathrm{T}} \in \mathscr{I}_2$ with $\mathscr{I}_1 = \{I_1 \in \mathbb{Z}^2 \mid 0 \le i_1 < 8, 0 \le j_1 < 8\}$, and $\mathscr{I}_2 = \{I_2 \in \mathbb{Z}^2 \mid i_2 = 0, 0 \le j_2 < 4\}$.

In Step (3) of our mapping flow according to Fig. 10.3, a *space–time mapping* (allocation and scheduling) of the transformed program needs to be carried out.

10.3.1.2. *Space–time mapping*

A *space–time mapping* assigns each iteration point $I \in \mathscr{I}$ to a processor (PE) index $p \in \mathscr{P}$ (*allocation*) and a time index $t \in \mathscr{T}$

(*scheduling*) may be obtained by the following transformation

$$\begin{pmatrix} p \\ t \end{pmatrix} = \begin{pmatrix} Q \\ \lambda \end{pmatrix} \cdot I, \tag{10.2}$$

where $I \in \mathscr{I}$, $Q \in \mathbb{Z}^{(n-1) \times n}$ and $\lambda \in \mathbb{Z}^{1 \times n}$. $\mathscr{T} \subset \mathbb{Z}$ is a set called *time space* that contains the set of start times t of all the iterations $I \in \mathscr{I}$. $\mathscr{P} \subset \mathbb{Z}^2$ is a set called *processor space* and the vector p *processor index*.

In the case of LSGP partitioning, all computations of iterations within a tile are executed in a pipelined way by the one processor assigned to the tile and multiple tiles may be executed in parallel as well by the different processors. The corresponding space–time mapping is given as follows:

$$\begin{pmatrix} p \\ t \end{pmatrix} = \begin{pmatrix} 0 & E \\ \lambda_1 & \lambda_2 \end{pmatrix} \cdot \begin{pmatrix} I_1 \\ I_2 \end{pmatrix}. \tag{10.3}$$

As can be seen, the processor space has the cardinality of $\mathscr{I}_2$, the number of tiles to cover all iterations. Moreover, the so-called *schedule vector* $\lambda = (\lambda_1 \ \lambda_2)$ has to be determined such that intra-tile as well as inter-tile data dependencies are satisfied.

Based on the above definitions, we may finally determine the start time $t(v_i, I)$ of each operation (= computation of the left-hand side variable $x_i[I]$) within the given loop body of iteration I as $t(v_i, I) = \lambda \cdot I + \tau(v_i)$ where $\tau(v_i)$ denotes an individual offset of the computation of the variable corresponding to v_i with respect to the beginning of variable computations belonging to loop iteration I.

10.3.1.3. *Scheduling, resource allocation and binding*

The purpose of *scheduling* is to determine a schedule vector λ according to Eq. (10.2) such that the *global latency* GL (total execution time) of a loop specification is minimized. As has been said, scheduling may also determine a *relative start time* $\tau(v_i)$ for each node v_i in the RDG corresponding to variable x_i in the program. The execution latency required for computing all variables x_i defined at an iteration I is called *local latency* LL. However, the time between starting the computation of two successive iterations assigned to a processor may be typically much smaller and is called *iteration interval II*. An iteration interval $II < LL$ therefore implies the (software) pipelining of iterations.

The purpose of *resource binding* is to assign to each node of the RDG a functional unit (resource) which computes the corresponding variable by evaluating the expression on the right-hand side of the corresponding loop specification.[d] The problem of optimally scheduling a loop nest under constraints of limited physical resources (PEs as well as functional units within each PE) is an instance of a resource-constrained scheduling and binding problem. It is solved here using a *mixed integer linear programming (MILP)* formulation [Ref. 13] and [Ref. 10]. Note that this approach obtains a global latency-optimal schedule for the entire loop nest instead of traditional modulo scheduling techniques [Ref. 5] which may schedule only the innermost loop. Additionally, beside the number of considered PEs and functional units within them, our approach encompasses also register and interconnect constraints. By solving the MILP using a state-of-the-art LP solver such as CPLEX [Ref. 17], both a *schedule vector* λ and the offset $\tau(v_i)$ of each node v_i are obtained, given an iteration interval *II*.

Example 10.4. For the running FIR filter, a feasible space–time mapping of the LSGP-partitioned iteration space is defined by Eq. (10.4).

$$\begin{pmatrix} p \\ t \end{pmatrix} = \begin{pmatrix} 0 & 0 & 1 & 0 \\ 0 & 0 & 0 & 1 \\ 8 & 1 & 0 & 8 \end{pmatrix} \cdot \begin{pmatrix} I_1 \\ I_2 \end{pmatrix} \quad \text{where } I_1 = \begin{pmatrix} i_1 \\ j_1 \end{pmatrix}, \ I_2 = \begin{pmatrix} i_2 \\ j_2 \end{pmatrix}.$$

$$(10.4)$$

An abstract view of the obtained processor array architecture for computing the $N = 32$ filter taps is shown in Fig. 10.5(a). The

---

[d]Note that in case this expression should be more complex than a unary or binary operator that is typically mappable to a single processor instruction, complex expressions may be split into simple equations by introducing intermediate variables that are assigned subexpressions with one or two operands only.

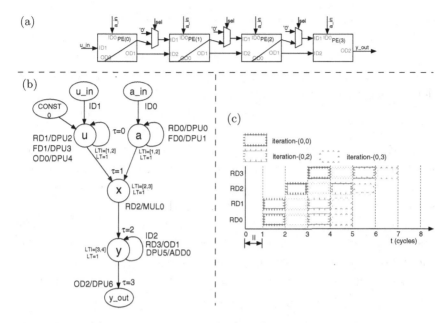

Fig. 10.5. In (a), abstract view of a 4 PE (1 × 4) processor array showing also the interconnection configuration between the shown PEs. In (b), RDG of the FIR filter specification introduced in Example 10.1. In (c), a possible register allocation for the first four consecutive loop iterations is shown using four registers.

resulting architecture is a 1×4 processor array and the corresponding processing elements are denoted $PE(0)$, $PE(1)$, $PE(2)$ and $PE(3)$, respectively. The arrows between the PEs in Fig. 10.5(a) show required communications of computed variable results.

In Fig. 10.5(b), a simplified version of the corresponding RDG is shown. After scheduling, the RDG is annotated with the local schedule (scheduling offsets) of nodes, denoted as τ (see Fig. 10.5(b)). From both schedule vector λ and offsets τ, the exact execution time of any operation (statement in the program) for each iteration becomes known. For instance, variable x at loop iteration $I_1 = (7\ 7)^{\mathrm{T}}$ ($I_2 = (0\ 0)^{\mathrm{T}}$) is computed by processor $PE(0)$ at time $\lambda_1 \cdot I_1 + \lambda_2 \cdot I_2 + \tau(\mathbf{x})$; i.e. at $(8\ 1) \cdot (7\ 7)^{\mathrm{T}} + (0\ 8) \cdot (0\ 0)^{\mathrm{T}} + 2 = 65$ cycles. Moreover, for the FIR filter as scheduled above, an iteration interval (*II*: the time in clock cycles between the evaluation of two successive iterations) of $II = 1$ cycle is assumed, and a local latency (*LL*: the time in clock cycles required to execute one iteration) of four cycles obtained on the basis of a given VLIW processor specification. The optimized total time to execute the FIR filter (the global latency) (GL) then evaluates to $GL = 8(T - 1) + 7 + 24 + 4 = 8T + 27$ cycles. It is determined by the interval between the start of the first iteration of $PE(0)$ and the last iteration being finished by $PE(3)$ which has the coordinates $I_1 = ((T - 1)\ 7)^T$.

During the solution of the resource-constrained scheduling problem, only the number of available functional units and registers is considered (allocation), whereas the binding is performed in a subsequent step. Since our considered CGRA architecture supports single cycle operations and thanks to functional pipelining, a new operation can start in each cycle in each functional unit. Moreover, the binding of scheduled operations to functional units becomes pretty simple. In contrast, the register binding is more difficult because variables may have have lifetimes of multiple cycles.

10.3.1.4. *Register binding*

Register binding is the process of assigning each program variable a register of the processor without violating any lifetime constraints. For each variable **x** of the given loop program, a *lifetime interval* LTI (the time interval beginning at the production of the variable LTI^{lb} and ending at the time LTI^{ub} where no more operations consume it) and the *lifetime* $(LT = LTI^{ub} - LTI^{lb})$ can be annotated in the RDG (see Fig. 10.5(b)). This information is necessary for the allocation of registers.

Example 10.5. For instance, let variable **a** of our scheduled FIR filter specification have a lifetime of $LT(a) = 1$. Hence, even for the minimal iteration interval of $II = 1$, the same register $RD0$ is used in each iteration as depicted in the Gantt chart in Fig. 10.5(c).

In our approach, register constraints within a processor may be considered as well. Here, our assumption is that enough registers are available to support a desired user-specified iteration interval II. Otherwise, scheduling will fail since register spilling, which is available in the case of standard processors, is not supported. Assigning proper input (ID), output (OD), general purpose registers (RD) and feedback data registers (FD) to variables is not a trivial task but can be achieved automatically based on the space–time mapping, dependence analysis, and using a modified left edge algorithm [Ref. 18]. For our running FIR filter specification introduced in Example 10.1, register binding is explained qualitatively below.

Example 10.6. (FIR-filter continued) The input variables a_in and u_in in the FIR filter program are assigned to input registers $ID0$ and $ID1$, respectively. Variable a is used to store the input a_in (filter coefficient) for reuse. However, it is assigned not only to one internal processor register $RD0$, but also to a feedback register $FD0$ for the following reasons:

- $RD0$: Every cycle, a new iteration is started according to the iteration interval $II = 1$. This implies that a new instance of all RDG operations will start executing at each cycle. In the first eight iterations and hence cycles, a new coefficient is read from the input $ID0$ (variable a_in). This is copied (assigned) to variable a. Here, $RD0$ is used to store variable a for PE-internal use within an iteration. To move the data from the input, functional unit $DPU0^{\mathrm{e}}$ is used.
- $FD0$: Feedback data registers are used to store variables that are cyclically reused. They are used to implement loop carried dependencies (dependencies spanning multiple iterations). For example, in case of processor $PE(0)$ is computing iteration $(i_1 \; j_1)^{\mathrm{T}} = (1 \; 0)^{\mathrm{T}}$, the value of a is the same as a at iteration $(i_1 \; j_1)^{\mathrm{T}} = (0 \; 0)^{\mathrm{T}}$ (due to localization for efficient data reuse); hence, when computing a at iteration $(i_1 \; j_1)^{\mathrm{T}} = (0 \; 0)^{\mathrm{T}}$, it has to be stored in a feedback data register so that it may be used again at the cycle where $(i_1 \; j_1)^{\mathrm{T}} = (1 \; 0)^{\mathrm{T}}$ is computed within the same processor. The length (adjustable delay in clock cycles) of the feedback data register may be determined easily by multiplying the intra-tile schedule vector λ_1 and the dependence vector d associated with variable a. In

---

^{e}A DPU is a functional unit that may move data from a source to a destination register.

our example, $\lambda_1 \cdot d_a = (8\ 1) \cdot \binom{1}{0} = 8$ (cf. Fig. 10.4). That is, eight coefficients of the FIR filter are read once from an input, and are then reused cyclically by reading from feedback register $FD0$.

Variable u is used to propagate the input filter input u_in. It is assigned to the registers $RD1$, $FD1$ for similar reasons as explained for variable a. However, u must even be assigned to a third register $OD0$ (output register).

- $OD0$: Due to partitioning, dependencies are split between local communication and data to be propagated to neighbor processors. For instance, $PE(1)$ must receive the filter inputs u from $PE(0)$ (see Fig. 10.4). Hence, in order to support this communication, variable u is also assigned to an output register $OD0$ that is physically connected to input register $ID1$ of the right neighbor processor, see also Fig. 10.5(a).

Variable x is assigned to register $RD2$ only as it is used internally only and only during the computation of an iteration. Finally, variable y is used to compute a partially accumulated sum of the filter output y_out that is output at the rightmost processor ($PE(3)$ in our example). y is assigned to register $RD3$. As this partial filter output is also communicated to the right neighbor processor, y is assigned also to register $OD1$.

Eventually, the functional unit and register binding is also annotated to the RDG (shown in Fig. 10.5(b)). From this tagged RDG, one could select finally the instructions for each node by looking at the incoming source nodes, i. e., assigned registers, the functional and register binding for that node. For instance, the instruction for node x in Fig. 10.5(b) would be MUL0 MUL RD2 RD1 RD0 in which MUL0 is the used functional unit, MUL is the mnemonic. RD2 is the destination operand, RD1 is source operand 1, and RD0 is source operand 2.

10.3.2. Back end

In this section, we give an overview of our back end mapping flow as depicted in Fig. 10.3. The back end generates (a) the configuration code for interconnection of processors and (b) the assembly code (programs) for the processors as explained below.

10.3.2.1. *Interconnect configuration*

In order to synthesize the configuration of the correct processor interconnection structure, the data dependencies of the PRA have to be analyzed. For this purpose, we introduce the terms *processor displacement* and *time displacement* by considering the following PRA equation and the space–time mapping given by Eq. (10.2).

$$x_i[I] = \mathscr{F}_i(\ldots, x_j[I - d_{ji}], \ldots) \text{ if } \mathscr{C}_i^I(I). \tag{10.5}$$

The synthesis of a processor interconnection for the data dependency d_{ji} from $x_j[I - d_{ji}] \to x_i[I]$ is done by first determining the *processor displacement* as follows:

$$d_{ji}^p = QI - Q(I - d_{ji}) = Qd_{ji}. \tag{10.6}$$

For each pair of processors p_t and $p_t = p_s + d_{ji}^p$ with $p_s, p_t \in \mathscr{P}$, a corresponding connection has to be configured from the source processor p_s to the target processor p_t. The *time displacement* denotes the number of time steps the value of variable $x_j[I - d_{ji}]$ must be stored before it is used to compute variable $x_i[I]$. Let v_i and v_j denote the corresponding RDG nodes for x_i and x_j. Then, given a schedule λ, the time displacement may be computed as follows:

$$
\begin{aligned}
d_{ji}^t &= \lambda I + \tau(v_i) - (\lambda(I - d_{ji}) + \tau(v_j) + w_j) \\
&= \lambda d_{ji} + \tau(v_i) - \tau(v_j) - w_j, \tag{10.7}
\end{aligned}
$$

where w_j denotes the computation time for computing variable x_j. In the processor array, the time displacement corresponds to the number of delay registers on the respective input port of the target processor p_t.

10.3.2.2. *Scalable compact code generation*

Obviously, code generation involves the generation of $|\mathscr{P}|$ binary codes, one for each processor of the CGRA architecture to be loaded and then executed in sync with the others. However, many processors will receive exactly the same program as (a) the schedule given is global, including iteration interval II and local latency LL. Hence, the code optimization· may not be performed independently for each sets of iterations to be assigned to an individual processor but must obey a global schedule of iterations. (b) Often, there are no iteration-dependent conditionals specified in the loop body. As a result, the computations to be performed are not only iteration independent but often even not different from processor to processor.

Unfortunately, our loops (PLA specifications) allow for iteration-dependent definitions of computations. Therefore, the number of different codes to generate may be more than one, but we will show nevertheless that through the definition of so-called *processor classes*, we may distinguish which minimal set of different cases do exist that are requiring to generate individual assembly codes. Informally, a so-called *processor class* is given by a unique combination of iteration-dependent conditions of loop variables to be computed. They are given implicitly by the possible intersections of iteration spaces of variables in the given loop nest.

Example 10.7. For the FIR filter shown in Fig. 10.4, three different processor classes may be distinguished. The leftmost processor $PE(0)$ distinguishes itself from others by reading in the filter inputs u_in from the left border of the array. Similarly, the rightmost processor $PE(3)$ is the only one to compute equation S9 (variable y_out) in addition to S8 (variable y). The processors assigned to the tiles in between all have the same set of statements to compute. Hence, we may have to generate only three types of codes, one for each processor class. Note that this number is independent of the problem size N of the filter specification.

We are therefore restricting ourselves next to compact code generation for each processor class individually. Here, based on a given scheduled loop program, *flat yet lengthy code* could be generated automatically by synthesizing assembly code for iteration after iteration to be executed. Even if allowing the overlap of the iteration execution in case $II < LL$(software pipelining), the code size would explode easily and require $LL + \det(P) \cdot II - 1$

VLIW words — making the generated code not only problem size or iteration space size dependent. Indeed, the code generated for one processor might not easily meet stringent constraints of tiny instruction memories inside each PE of a CGRA.

Moreover, in the case of our FIR filter specification, imagine that the specified number of samples T might not even be finite. In that case, the technique above cannot be used at all as $\det(P) = T \times N/PE_{num}$ also becomes infinite with PE_{num} denoting the number of processors.

In the following, we therefore address the problem of *code compaction* by (a) identifying and grouping together code sequences that are executed repeatedly through the notion of so-called *program blocks (PB)* and (b) generating as compact as possible code for a considered processor class by looping between these unique blocks at provably zero overhead (no additional cycles) so to preserve the given schedule λ.

Definition 10.3. (Program block) A *program block* is a unique sequence of assembly instructions whose length amounts exactly the local latency LL. Sets of such iterations characterized by the same sequence of instructions (same program block) may be described by an iteration space called *program block iteration space* $\mathscr{I}_{PB_i}$. Each iteration within this space requires the execution of the same instruction sequence. All program block iteration spaces $\mathscr{I}_{PB_i}$ partition the tile iteration space assigned to one processor: $\mathscr{I}_1 = \mathscr{I}_{PB_0} \cup \mathscr{I}_{PB_1} \cup \ldots \cup \mathscr{I}_{PB_{m-1}}$ and $\mathscr{I}_{PB_i} \cap \mathscr{I}_{PB_j} = \varnothing$ for all $0 \leq i,j, \leq m - 1$, $i \neq j$.

We explain the notion of program blocks for processor class 1 ($PE(1)$ and $PE(2)$) with the following example. A similar explanation is valid for the other two processor classes.

Example 10.8. $PE(1)$ of the CGRA shown in Fig. 10.5(a) has to execute six different program blocks as shown in Fig. 10.6(b), differentiated by different rectangles and is denoted by $PB0$, $PB1$, $PB2$, $PB3$, $PB4$, and $PB5$. For each program block, its RDG,

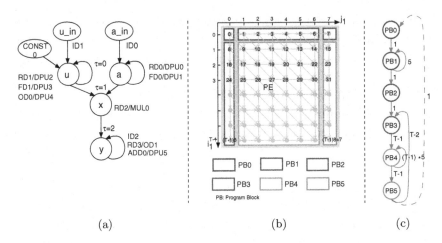

(a)   (b)   (c)

Fig. 10.6.   (a) RDG for the processor class of the middle processors ($PE(1)$, $PE(2)$) of the FIR filter. In (b), partition of the iterations of the tile specific for this processor class into program blocks. The numbers annotated to the iterations show the start time of the each iteration based on a given schedule λ_1. From the schedule, the program block control flow graph shown in (c) may be extracted. With one node per program block, the edges indicate the traversal of control flow based on a scanning of the tile iteration space as prescribed by the schedule. The edge weights indicate how often the corresponding transition between two program blocks is taken.

program block iteration space $\mathscr{I}_{PB_i}$, and its $LL = 4$ cycle VLIW assembly code can be seen in Fig. 10.7.

One notable difference between $PB0$ and $PB1$ is that although y in both program blocks is defined by equation S8 (y=x+y), in $PB0$, one of the input is $ID2$ whereas the same input in $PB1$ comes from an internal register $RD3$. The assembly level code for computing of y would be ADD RD3 RD2 ID2 in case of $PB0$ and ADD RD3 RD2 RD3 in case of $PB1$, respectively. Moreover, in $PB0$, u_in is accessed from an input port, but in $PB1$, u_in is always zero (constant), i.e. they differ in the instruction where u_in is read.

The difference between $PB1$ and $PB2$ is that in $PB1$, the partial sum (y) is written to the internal data register of the processor (ADD RD3 RD2 RD3). In $PB2$, the partial sum is written to the output register of the processor (ADD OD1 RD2 RD3) for transfer to the right neighbor processor. Hence, they will differ in at least one instruction.

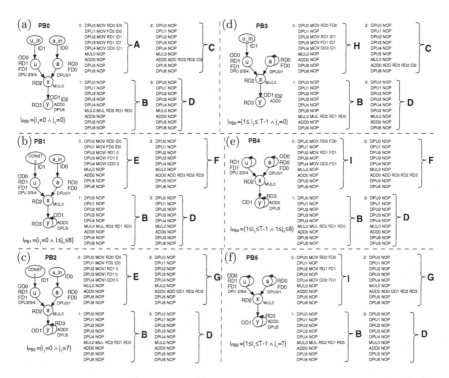

Fig. 10.7. Program blocks of $PE(1)$ of the FIR filter shown in Fig. 10.5(a). (a) RDG of $PB0$, program block iteration space, and naive code. Since the local latency $LL = 4$, each program block consists of 4 VLIW instructions denoted as 0, 1, 2 and 3, which may contain NOPs as well. Similarly, respective RDGs, iteration spaces and their codes for program block $PB1$ in (b), $PB2$ in (c), $PB3$ in (d), $PB4$ in (e), and $PB5$ in (f) are shown. Different unique VLIW instructions from all program blocks are labeled as A,B,C,D,... to differentiate easily from one another.

For variable a in $PB3$, its value has to be read from a feedback data register instead of an input port (in case of $PB0$). In $PB4$, both u and a values have to be read from feedback shift registers as there is no direct access to inputs. Moreover, in $PB5$, variable y has to be written to an output port in contrast to $PB3$ and $PB4$; thus, $PB3$, $PB4$, and $PB5$ differ in instructions from each other and from $PB0$, $PB1$, and $PB2$, respectively (see Fig. 10.7). As a result, the iteration space is partitioned into a total of six different segments of code.

Putting it all together, given a nested loop specification, its iteration space gets partitioned into as many tiles as available CGRA processors during partitioning. For each identified processor class, a code compaction problem has to be solved preserving a given schedule of iterations, functional unit and register assignment. Each tile belonging to a processor class itself consists of different program blocks that describe unique sequences of instructions to execute one iteration of the partitioned loop program. In the following, we describe the process of code compaction based on a characterization of program blocks.

10.4. Compact Code Generation

Obviously, the execution of iterations in a tile assigned to one processor shall follow a given schedule. Note that in typical cases where the local latency of a program block LL is longer than the iteration interval II, corresponding iteration executions do overlap (software pipelining). Given this understanding, we explain in the following how to generate the most compact code for such a given schedule of iterations by not generating flat code for each iteration but by exploiting repetitive sequences of program block executions by the introduction of looping between program blocks under the assumption that such introduction of branches between program block codes does not create any additional execution cycles (zero overhead looping).

Since each iteration uniquely belongs to one program block, the program execution according to a given schedule can be represented also visually as a traversal of a program block graph in which nodes correspond to program blocks.[f] Based on this observation, the procedure of generation of compact code mainly involves the following three steps:

(1) Generation of a *program block control flow graph* which represents the execution order of successive program block executions.

---

[f]Keep in mind that their executions may, however, overlap.

(2) Minimally unfolding of cycles in this graph which may be required to reflect restrictions caused by register binding and variable lifetimes.

(3) Generation of assembly code once for each block including the insertion of proper branching subwords at block boundaries to guarantee a correct schedule of program block executions. These have to be scheduled inside a processor's branch unit at zero time overhead so to preserve the throughput and latency of the given schedule.

In a *program block control flow graph* (G_{CF}, see Fig. 10.6(c)), each node corresponds to a program block. The number annotated to each edge indicates how many times the execution is transferred from the predecessor to the successor program block in the overall traversal of the iteration space assigned to the respective processor.

10.4.1. *Generation of program block control flow graph*

In this section, we present an algorithm to generate the program block control flow graph G_{CF} on whose basis CGRA code is emitted. As each processor must execute the iterations assigned to it in a *scanning order* as imposed by the given intra-tile schedule λ_1, we introduce the notation of a *path stride matrix* to simplify the computation of G_{CF} as follows.

Definition 10.4. (Path stride matrix) The path strides of an n-dimensional parallelotope may be represented by a matrix $S \in \mathbb{Z}^{n \times n}$ consisting of n (column) vectors $\vec{s_i} \in \mathbb{Z}^n$ and given as $S = (\vec{s_1} \, \vec{s_2} \, \ldots \, \vec{s_n})$. The path stride matrix defines the scanning (execution order) of iterations within a tile.

Example 10.9. For the LSGP-partitioned FIR filter shown in Fig. 10.4, the path stride matrix respecting the *intra-tile* schedule given by $\lambda_1 = (8 \, 1)$ is $S = \left(\begin{smallmatrix} 0 & 1 \\ 1 & -7 \end{smallmatrix} \right)$. Each tile is processed by a processor in the order imposed by S. For $PE(1)$ in our example according to Fig. 10.6(b), the iteration space is executed starting at

$(i_1\ j_1)^{\mathrm{T}} = (0\ 0)^{\mathrm{T}}$ to $(i_1\ j_1)^{\mathrm{T}} = (0\ 7)^{\mathrm{T}}$, scanning in the direction of $\vec{s_1} = (0\ 1)^{\mathrm{T}}$ until the tile end is reached. Then, adding $\vec{s_2} = (1\ -7)^{\mathrm{T}}$, the next iteration scanned is $(i_1\ j_1)^{\mathrm{T}} = (1\ 0)^{\mathrm{T}}$ and executed. Then again, the scanning in the direction of $\vec{s_1}$ continues, and so on.

In the following, we assume that the schedule λ_1 has been determined so to minimize the global latency GL for a given throughput (iteration interval II) and that the corresponding stride matrix S reflecting this schedule has been constructed accordingly.

Algorithm 10.1 Generation of Program Block Control Flow Graph (G_{CF}).

Input: Set $\mathbf{PB} = \{\mathscr{I}_{PB_0}, \ldots, \mathscr{I}_{PB_{m-1}}\}$ of *PBs*, path stride matrix S.

Output: $G_{CF}(V, E, w)$
    // V : node set; E : edge set; w(e), e ∈ E : edge weight
    $V = \{v_0, \ldots, v_{m-1}\}$;   *//graph initialization; one node per PB*
    $E = \emptyset$; *// no edges*
    for each $i \in \{0, \ldots, m-1\}$ **do**
      for each $\vec{s} \in \vec{S}$ **do**
        //for each stride vector $\vec{s}$
        $\mathscr{I}_{\vec{s}} = \{I + \vec{s} \mid \mathscr{I} \in \mathscr{I}_{PB_i}\}$ *//$\mathscr{I}_{PB_i}$ is the iteration space of*
        PB_i
        //$\mathscr{I}_{\vec{s}}$ is the space translated by $\vec{s}$
        //This space is intersected with all other iteration
          spaces
      for each $k \in \{0, \ldots, m-1\}$ **do**
        if $(\mathscr{I}_{\vec{s}} \cap \mathscr{I}_{PB_k} \neq \varnothing)$ **then**
        $w = |\mathscr{I}_{\vec{s}} \cap \mathscr{I}_{PB_k}|$;
        if $(w > 0)$ **then**
          $E = E \cup \{(v_i, v_k)\}$;
          $w((v_i, v_k)) = w$;

We propose now Algorithm 10.1 that constructs the program block control flow graph for a given processor class according to a given schedule λ_1, respectively corresponding stride matrix S. The graph contains as many nodes as program blocks. In order to determine which transitions do occur between pairs of program blocks, we need to find out whether it is possible and how often the scanning of the tile space involves a transition from one to another block. This is achieved in Algorithm 10.1 by checking for each program block PB_i whether the space obtained when adding each time a column vector $\vec{s}$ of the stride matrix S to its iteration space $\mathscr{I}_{PB_i}$ has a nonzero intersection with the iteration space of other program blocks, including its own. If the intersection is empty, there is no transition between the program blocks. Else, there is a transition between them and an edge is inserted between the corresponding two program block nodes. Moreover, by counting the number of points in each intersection, the corresponding edge weights denoting the number of times the execution is transferred from one program block to the other may be also determined and annotated to the edges of the graph G_{CF}.

Example 10.10. For the FIR filter shown in Fig. 10.4, processor $PE(1)$ is shown in Fig. 10.6(b) and its G_{CF} in Fig. 10.6(c). Here, the execution enters $PB0$, executes corresponding iteration, then enters $PB1$ and executes six times the code of this block consecutively. Then, the control flow enters $PB2$, executing one iteration. Afterwards, the execution enters $PB3$, again executing one iteration, then executes six times $PB4$ consecutively before entering and execution of the one iteration belonging to $PB5$. Finally, the execution follows the sequence $PB3 \rightarrow 6 \times PB4 \rightarrow PB5$ again and again.

10.4.2. *Graph transformation and final code emission*

Before the emission of final code, cycles in the control flow graph G_{CF} might need to be partially unfolded first in order to satisfy given register binding constraints [Ref. 19]. More specifically, instructions from different consecutively executed iterations belonging to a program block (node in G_{CF}) are arranged contiguously according to the schedule in order to form so-called *overlapped codes*. Afterwards, these overlapped codes of a program block have to be combined with all direct successor nodes in the G_{CF}. Although we are using the term overlapped code, note that only in the last step "real" code is emitted.

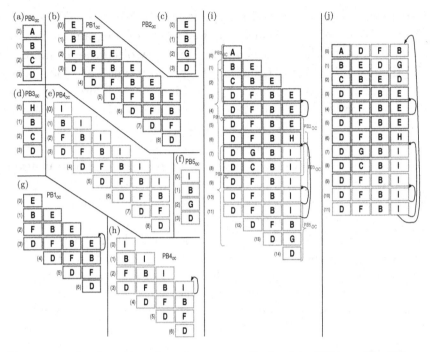

Fig. 10.8. Overlapped codes of program blocks (a) $PB0_{OC}$, (b) $PB1_{OC}$, (c) $PB2_{OC}$, (d) $PB3_{OC}$, (e) $PB4_{OC}$, (f) $PB5_{OC}$ for the partitioned and scheduled FIR filter example and for the processor class of middle processors ($PE(0)$, $PE(1)$). Note that in (a) there is only one iteration in this program block. Thus the overlapped code for this program block is same as for non-overlapped code. (g) Optimized overlapped code of $PB1$, $PB1_{OC}$. (h) Optimized overlapped code of $PB4$, $PB4_{OC}$. (i) Resulting overlapped codes from step two, i.e. combining of overlapped codes from different program blocks. (j) Final optimized code consisting of 12 VLIW instruction words. The arcs annotated to the code sequence denote run-time selected branch targets.

Example 10.11. An example of overlapped codes of the program blocks is shown in Fig. 10.8. The overlapped code of $PB0$, denoted as $PB0_{OC}$ is shown in Fig. 10.8(a). There is only one iteration that gets executed when the execution enters $PB0$; hence there is no possibility for contiguous arrangement of instructions from different iterations. Whereas for $PB1$, the overlapped code ($PB1_{OC}$) is shown in Fig. 10.8(b) in which instructions from six different iterations are arranged contiguously since six iterations are executed consecutively whenever the execution enters $PB1$. Similarly, overlapped codes of other program blocks $PB2$, $PB3$, $PB4$, and $PB5$, denoted by $PB2_{OC}$, $PB3_{OC}$, $PB4_{OC}$, and $PB5_{OC}$, are shown in Figs. 10.8(c), (d), (e), and (f).

A closer look at the overlapped code of $PB1$, $PB1_{OC}$ shown in Fig. 10.8(b) reveals that there is redundancy in the instructions. Particularly, the instructions (4) and (5) are the same as instruction (3). The overlapped code for $PB1$ can be further optimized as shown in Fig. 10.8(g). In this figure, a branch (jump) instruction denoted by an arc is introduced from instruction (3) to (3). It is the responsibility of the controller to generate appropriate control signals so that this branching happens at the appropriate iteration. However, introducing such a branch instruction does not affect the other data flow operations and preserves the given schedule. Afterwards, these optimized overlapped codes of program blocks are arranged contiguously again with respect to the given schedule and G_{CF} as shown in Fig. 10.8(i).

For the code in Fig. 10.8(i), there is still a further possibility to reduce the code size by folding the *epilogue* [Ref. 20] instructions into the *prologue* instructions as shown in Fig. 10.8(j). However, this would add an extra branch/jump instruction from instruction 11 to instruction 0. In case of the running FIR filter example, our approach finally emits a *final optimized code (FOC)* of 12 VLIW instructions only as shown in Fig. 10.8(j) whereas flat and lengthy code would require $4 + (T \times N/PE_{num}) - 1$ VLIW instructions which would evaluate to 67 already for just $T = 8$ samples ($N = 32$ filter taps, $PE_{num} = 4$ processors). It can be verified easily that the code size will remain constant and independent of the size of the iteration space as long as schedule and allocation remain the same. Therefore, this code is as compact as possible and at the same time iteration-space-independent. For instance, either increasing the number of filter taps N or the number of input samples T will only change the configuration parameters of the controller, but not the length of the assembly codes of the processors.

10.4.3. *Controller*

In the considered CGRA architecture, the reconfigurable *controller* (see Fig. 10.1) plays an important role: the entire static loop control flow (i.e. the branching between program blocks) of the given nested loop program is captured by this controller. The purpose of it is to scan the entire iteration space according to a given schedule and to issue a set of control signals to the processors each cycle encoding the information when to jump and to which next program block segment. This determines basically the static control flow within the programs of each PE. The control signals values thereby depend on the current iteration I and the program block it belongs to as shown in the following Algorithm 10.2. According to this generic algorithm, the number of control signals needed is equal to the number m of program blocks (one-hot coding of each binary control signal issued with a program block). In the branching unit of each PE, these control signals are received and decoded in parallel. A decision is taken in the same cycle whether and where to jump, i.e. which program block to execute next.

Algorithm 10.2 Generic controller algorithm generating branching control signals for the PEs.

Input: Iteration space $\mathscr{I}$ and program block iteration spaces $(\mathscr{I}_{PB_0}, \ldots, \mathscr{I}_{PB_{m-1}})$

Output: Control signals $(IC_0, \ldots, IC_{m-1})$

    for $i_1 = lb_1$; $i_1 \leq ub_1$; $i_1 = i_1 + 1$ **do**

        for $i_2 = f_{1_{lb}}(i_1)$; $i_2 \leq f_{1_{ub}}(i_1)$; $i_2 = i_2 + 1$ **do**

          $\vdots$

            for $i_n = f_{n_{lb}}(i_1, \ldots i_{n-1})$; $i_n \leq f_{1_{ub}}(i_1, \ldots i_{n-1})$; $i_n = i_n + 1$ **do**

                //next iteration is $I = (i_1 \ i_2 \ \ldots \ i_n)^{\mathrm{T}}$

                if $(I \in \mathscr{I}_{PB_0})$ **then** $IC_0 = 1$ **else** $IC_0 = 0$

                if $(I \in \mathscr{I}_{PB_1})$ **then** $IC_1 = 1$ **else** $IC_1 = 0$

                $\vdots$

                if $\left(I \in \mathscr{I}_{PB_{m-1}}\right)$ **then** $IC_{m-1} = 1$ **else** $IC_{m-1} = 0$

In the hardware implementation of the control algorithm, registers are allocated and initialized first with the iteration variables and their bounds. Depending on schedule and iteration interval, appropriate iteration variables need to be incremented/decremented in each cycle so that the current iteration vector (I) and the program block to which it belongs are known and proper control signals determined. Note that the hardware overhead of such a programmable controller unit implementation occurs only once. However, the implementation imposes a single cycle computation and emission requirement to compute the bundle of control signals denoted $IC_{0,1,...,m-1}$.

In the future, we will show that through optimization and when using a more efficient encoding of binary control signals, the number of control signals may be greatly reduced to the binary logarithm of the maximum number of outgoing edges of a program block in the program block control flow graph, see, e.g. Fig. 10.6(c). This approach will also reduce the hardware cost of the controller.

Example 10.12. In the continuation of our FIR filter example, consider instruction (10) in Fig. 10.8(j). When this instruction is executed, there are two possibilities for the next instruction to be executed. The execution must either jump to instruction (10) again if the next iteration vector I belongs to iterations of program block $PB4$ again ($1 \leq i_1 \leq T - 1, 1 \leq j_1 \leq 6$). Elsewise, program block $PB5$ needs to be executed ($1 \leq i_1 \leq T - 1, j_1 = 7$). Obviously, the control decision which way to branch is dependent only on the separating hyperplane $j_1 = 6$ resulting in the following segment of control code that will be executed in hardware, producing one set of control signals every II cycles.

```
    . . .
    for (i₁ = 0, i₁ ≤ T − 1, i₁ = i₁ + 1)
    { for (j₁ = 0, j₁ ≤ 7, j₁ = j₁ + 1)
      {
          . . .
          if (j₁ == 6) then ICₓ = 1 else ICₓ = 0;
          . . .
      }
    . . .
    }
```

IC_x is a binary control signal which is decoded by the branch unit as part of the execution of instruction (10) as follows: IF ICx, JMP 11, JMP 10. If $IC_x = 1$, the execution will proceed with the next instruction (11), else it will repeat instruction (10) again.

10.5. Results and Discussion

Our approach for compact code generation has been integrated into an existing loop compiler [Ref. 14], written in C++. As a CGRA platform, a processor array of VLIW cores with reconfigurable interconnect is selected and configured. In a first set of evaluations, the scalability of our proposed code generation technique for different problem sizes and algorithms is studied, see Table 10.1.

In the first set of experiments, the FIR filter latency is explored for different numbers N of filter taps (problem size) assuming a constant number of processors PE_{num} (1-D processor array of size $1 \times PE_{num}$). Here, it can be seen that if the problem size N increases, the latency increases but the code size remains the same. The primary reason for this behavior is that if the problem size increases, the number

Table 10.1. Scalability and code size independence for different applications for variable (a) problem size and (b) CGRA processor number PE_{num}.

Application	PE_{num}	Global Latency (#cycles)	code size (#instr./PE)
FIR $(N = 32, T = 32)$	1	1027	12
FIR $(N = 32, T = 32)$	2	531	12
FIR $(N = 32, T = 32)$	4	283	12
FIR $(N = 64, T = 32)$	1	2051	12
FIR $(N = 64, T = 32)$	2	1059	12
FIR $(N = 64, T = 32)$	4	563	12
FIR $(N = 64, T = 32)$	8	315	12
MM (16×16)	1×1	4099	16
MM (16×16)	2×2	1047	16
MM (32×32)	1×1	32771	16
MM (32×32)	2×2	8231	16
MM (32×32)	4×4	2111	16

of program blocks remains the same — however, program blocks may need to be repeated more often, which is orchestrated by the controller. In other words, the number of iterations that are assigned and executed by a processor is increased but not at the expense of an increased code size. Second, it can also be seen how a tradeoff in global latency GL may be achieved depending on the number of processors PE_{num} available in the CGRA. The same experiment is carried out for a matrix multiplication example being mapped to a 2-D array configuration of different problem and array sizes.

In the second experiment, we consider again the FIR filter application, a matrix multiplication (MM), and in addition a sum of absolute differences (SAD) application stemming from H.264 video coding. The generated codes were simulated on a cycle-accurate simulator. In the following experiment, our code generation approach is tested and compared for achievable throughput, code size and overheads for single-processor targets such as the HPL-PD architecture and the Trimaran compilation and simulation infrastructure [Ref. 21].[g]

In Trimaran, modulo scheduling [Ref. 5] as well as rotating register files were used. Since our CGRA architecture is a streaming architecture, special address generators (see Fig. 10.1) that are responsible for fetching the correct data from the input buffers and writing the results back to the output buffers, are needed at the borders of the array. In case of Trimaran, only load/store units are needed. In order to match our test environment with Trimaran, we have modified the HPL-PD architecture description in the following way: we specified three memory units as all the applications that we considered need two inputs and one output. Also, to make a fair comparison, we have specified three extra adders (integer units) in Trimaran for address generation and for loop counter incrementation. Finally, all instructions are assumed to be single cycle operations in the HPL-PD architecture as well as in our architecture. With these

---

[g]Hence, the iteration space of the whole loop program is partitioned into a single tile only to reflect that all iterations will be assigned to a single processor (LSGP partitioning used).

Table 10.2.   Single processor comparison of proposed approach with Trimaran [Ref. 11]. A $N = 32$ tap FIR filter and a matrix-matrix multiplication (MM) are analyzed for two matrices of size 16×16. Finally, a SAD application is analyzed and compared based on a block size of 16×16 with respect of achieved average iteration interval, code length and overhead.

Application	Average Iteration Interval (II_{avg})		Number of VLIW instructions		Overhead (%)	
	Trimaran	Ours	Trimaran	Ours	Trimaran	Ours
FIR	2.28	1	13	12	14.06	0
MM	2.53	1	17	16	26.36	0
SAD	8.38	1	14	5	4.60	0

modifications, the Trimaran environment matches well with our test environment.

In Table 10.2, (a) the average iteration interval[h] (II_{avg}), (b) the number of VLIW instructions, and (c) the overheads[i] for different applications of our approach are compared against Trimaran. Our experimental results reveal the following facts:

(1) The average iteration interval II_{avg} is reduced in our approach by up to 88% over Trimaran which means $8.4 \times$ higher throughput. This is mainly due to the controller which generates all branching control signals for loop bound control without any timing overhead. The same configurable controller generates the control signals for all processor classes. Therefore, the cost of this controller is incurred only once. The extra hardware cost of the controller is almost compensated by extra comparators (for loop bound checking) and adders (for loop index incrementation) that are needed in Trimaran. The global latency is always less in our approach as we are able to achieve a better II_{avg}.

---

[h]The average iteration interval II_{avg} is the average time between the start of two successive loop iterations. It is calculated by dividing the total execution time of a loop nest by the total number of iterations executed.
[i]The overhead is evaluated as the amount of time that is spent in executing other than the innermost loop body compared to total execution time.

(2) The number of VLIW instructions is greatly reduced. This is mainly due to the separation of the control flow from the data flow in our considered architecture.

(3) Table 10.2 shows the overheads in Trimaran for different applications. In our approach, the overhead is always zero whereas in Trimaran the total execution time may increase up to 26.36% for the set of considered examples.

Despite the little overhead of the controller and address generators in hardware, our approach is promising since it scales very well with the size of the iteration space of a loop program and the number of instructions needed for a given arbitrarily nested loop application is always less compared with Trimaran. In summary, we are able to achieve a much smaller average iteration interval II_{avg} (higher throughput), see Fig. 10.9 [Ref. 11].

10.6. Related Work

Often, there is only a fine line between on-chip processor arrays and CGRAs since the provided functionality is similar. We refer to [Ref. 22] for an excellent overview on CGRAs. Even though there exists only little research work that deals with the compilation to CGRAs, we want to distinguish ours from it. The authors in [Ref. 23] describe a compiler framework to analyze SA-C programs, perform optimizations, and automatically map applications onto the MorphoSys architecture [Ref. 24], a row-parallel or column-parallel SIMD architecture. This approach is limited in the sense that the

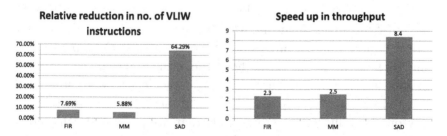

Fig. 10.9.   Relative reduction in terms of VLIW instructions and throughput for applications FIR, MM, and SAD compared to Trimaran.

order of the synthesis is predefined by the loop order and no data dependencies between iterations are allowed. Another approach for mapping loop programs onto CGRAs is presented by Dutt and others [Ref. 25]. Remarkable in their compilation flow is the target architecture, the *Dynamically Reconfigurable ALU Array*, a generic reconfigurable architecture template, which can represent a wide range of CGRAs. The mapping technique itself is based on loop pipelining and partitioning of a data flow graph into clusters, which can be placed on a line of the array. However, all the aforementioned mapping approaches follow a "traditional" compilation approach where first transformations such as loop unrolling are performed, and subsequently the intermediate representation in the form of a control/data flow graph is mapped by using placement and routing algorithms. Unique to our approach is that, thanks to using loop tiling in the polyhedron model [Ref. 12], the placement and routing is implicitly given — i.e. for free and much more regular.

In embedded applications such as signal processing and multimedia, most of the execution time is spent in nested loops. However, no loop execution without any loop control (e.g. loop bound condition checking, incrementing, and branching). This control overhead can be minimized by employing zero-overhead looping schemes [Ref. 26], which are available in most modern DSPs for innermost loops. In [Ref. 27], Kroupis *et al.*, proposed a single processor controller and compilation technique which not only maps the innermost loop control but also the outer loop control onto a controller realized in hardware so to support multi-dimensional zero loop overhead.

Our approach for mapping loops onto a processor array is different from [Ref. 27] in the following ways: first, we consider multiprocessor arrays and not single processors. Second, we strictly separate the control flow from the data flow (i.e. scheduling is as tight as possible and restricted only by data dependencies and functional unit and register resource constraints). Branching is performed based only on the values of control signals that are generated completely outside a processor (propagated from the controller located at the border of the CGRA). This separation of control flow is unique to our proposed code generation approach and finally also more energy-efficient as control signals needed for branching are computed only once and are propagated in a delayed fashion unmodified to the proper processors. Furthermore, our scheduling and mapping approach is able to exploit both instruction and loop level parallelism, respectively pipelining of iterations (software pipelining). As this may typically result in increased code sizes, we proposed a systematic way of code compaction based on the idea to compress naive loop code by exploiting repetitive segments of code called program blocks and multi-dimensional zero-overhead looping.

10.7. Conclusions

We presented an approach for compact code generation and optimization for a class of CGRAs. For the fast and energy-efficient execution of many types of nested loop programs, the proposed CGRA offers many configuration possibilities at the level of VLIW core specification and also a flexible topology configurability. Unique in the processor architecture specification are also reconfigurable feedback registers. Systems with 100s of cores are, however, limited in instruction and data memory available per processor. We therefore presented a methodology for compact code generation for large scale CGRAs starting with a quite general class of loop programs. Our main results are that typically, code does not need to be generated for each processor separately, but only for a small number of so-called *processor classes* that distinguish themselves by different functionality. Hence, each processor belonging to the same processor class will receive the same program. The program loading may be achieved in parallel using a *multi-cast configuration* method presented in [Ref. 6]. Subsequently, in order to avoid the explosion of code for a single processor, repetitive patterns of instruction sequences are extracted called *program blocks*. Moreover, the presented code generation and optimization methods based on program blocks are able to emit a most compact assembly code by preserving a given schedule of iterations and branching between the program block codes emitted only once instead of being replicated. The final code has been shown to be independent of the loop problem size and also size of the target CGRA.

Our approach shows promising results also compared to the Trimaran compilation infrastructure. It may be used also for fine-tuning an application with respect to the instruction memory requirements, hence can be used also during design space exploration [Ref. 28].

In the future, we would like to investigate symbolic code generation techniques [Refs. 20 and 30], which are especially beneficial when the number of available processing elements in a CGRA becomes known only at run-time [Ref. 31].

10.8. Acknowledgment

This contribution is dedicated to Professor Peter Y.K. Cheung at the occasion of his 60th birthday and in recognition of his outstanding contributions to the field of reconfigurable computing in general and field-programmable gate arrays (FPGA) in particular. The first author has followed the pioneering work of Professor Cheung already since being a PhD student more than 20 years ago. Being constantly fascinated by his inspiring thoughts and with admiration of his steady success in solving fundamental problems that have advanced the field of reconfigurable computing tremendously, we are very proud to congratulate with this chapter of our current research.

This work was supported by the German Research Foundation (DFG) as part of the Transregional Collaborative Research Centre "Invasive Computing" (SFB/TR 89).

References

1. V. Baumgarte *et al.* PACT XPP — A Self-Reconfigurable Data Processing Architecture, *The Journal of Supercomputing*, 26(2), 167–184, 2003.
2. F. Bouwens *et al.* Architecture Enhancements for the ADRES Coarse-grained Reconfigurable Array, in *Proc. International Conference on High Performance Embedded Architectures and Compilers*, pp. 66–81, 2008.
3. A. Duller, G. Panesar, and D. Towner. Parallel Processing — the PicoChip Way!, *Communicating Process Architectures*, pp. 125–138, 2003.

4. A. Olofsson. A 1024-core 70 GFLOP/W Floating Point Manycore Micro-processor, in *Proc. Annual Workshop on High Performance Embedded Computing*, 2011.

5. B.R. Rau. Iterative Modulo Scheduling: an Algorithm for Software Pipelining Loops, in *Proc. International Symposium on Microarchitecture*, pp. 63–74, 1994.

6. D. Kissler *et al.* A Dynamically Reconfigurable Weakly Programmable Processor Array Architecture Template, in *Proc. International Workshop on Reconfigurable Communication Centric System-on-Chips*, pp. 31–37, 2006.

7. J. Teich, L. Thiele, and L. Zhang. Scheduling of Partitioned Regular Algorithms on Processor Arrays with Constrained Resources, *Journal of VLSI Signal Processing*, 17(1), 5–20, 1997.

8. H. Dutta, F. Hannig, and J. Teich. Hierarchical Partitioning for Piecewise Linear Algorithms, in *Proc. International Conference on Parallel Computing in Electrical Engineering*, pp. 153–160, 2006.

9. J. Teich and L. Thiele. Exact Partitioning of Affine Dependence Algorithms, in *Proc. Embedded Processor Design Challenges*, pp. 135–153, 2002.

10. F. Hannig. Scheduling Techniques for High-Throughput Loop Accelerators, Dissertation, University of Erlangen-Nuremberg, Germany, 2009.

11. S. Boppu, F. Hannig, and J. Teich. Loop Program Mapping and Compact Code Generation for Programmable Hardware Accelerators, in *Proc. International Conference on Application-specific Systems, Architectures and Processors*, pp. 10–17, 2013.

12. P. Feautrier and C. Lengauer. "Polyhedron Model", In ed. D. Padua, *Encyclopedia of Parallel Computing*, pp. 1581–1592, Springer, New York, NY, USA, 2011.

13. F. Hannig and J. Teich. Resource Constrained and Speculative Scheduling of an Algorithm Class with Run-Time Dependent Conditionals, in *Proc. International Conference on Application-specific Systems, Architectures, and Processors*, pp. 17–27, 2004.

14. F. Hannig *et al.* PARO: Synthesis of Hardware Accelerators for Multi-Dimensional Dataflow-Intensive Applications, in *Proc. International Workshop on Applied Reconfigurable Computing*, pp. 287–293, 2008.

15. L. Thiele and V. Roychowdhury. Systematic Design of Local Processor Arrays for Numerical Algorithms, in *Proc. International Workshop on Algorithms and Parallel VLSI Architectures*, pp. 329–339, 1991.

16. M. Wolfe. *High Performance Compilers for Parallel Computing*, Addison-Wesley, Boston, MA, USA, 1996.

17. ILOG, CPLEX Division. *ILOG CPLEX 12.1, User's Manual*, 2011.

18. F. Wu *et al.* Simultaneous Functional Units and Register Allocation Based Power Management for High-level Synthesis of Data-intensive Applications, in *Proc. International Conference on Communications, Circuits and Systems*, 2010.

19. S. Boppu, F. Hannig, and J. Teich. Loop Program Mapping and Compact Code Generation for Programmable Hardware Accelerators, in *Proc.*

International Conference on Application-Specific Systems, Architectures and Processors, pp. 10–17, 2013.

20. B.R. Rau, M.S. Schlansker, and P.P. Tirumalai. Code generation schema for modulo scheduled loops, *ACM SIGMICRO Newsletter*, 23(1–2), 158–169, 1992.

21. Trimaran. An Infrastructure for Research in Backend Compilation and Architecture Exploration. [Online] Available at: http://www.trimaran.org/. [Accessed 23 April 2014].

22. T. Todman *et al.* Reconfigurable Computing: Architectures and Design Methods, *IEE Proc. Computers and Digital Techniques*, 152(2), 193–207, 2005.

23. G. Venkataramani *et al.* Automatic Compilation to a Coarse-grained Reconfigurable System-on-Chip, *ACM Transactions on Embedded Computing Systems*, 2(4), 560–589, 2003.

24. H. Singh *et al.* MorphoSys: An Integrated Reconfigurable System for Data-Parallel and Computation-Intensive Applications, *IEEE Transactions on Computers*, 49(5), 465–481, 2000.

25. J. Lee, K. Choi, and N. Dutt. An Algorithm for Mapping Loops onto Coarse-Grained Reconfigurable Architectures, *in Proc. Conference on Languages, Compilers, and Tools for Embedded Systems*, pp. 183–188, 2003.

26. G.-R. Uh *et al.* Effective Exploitation of a Zero Overhead Loop Buffer, *in Proc. Conference on Languages, Compilers, and Tools for Embedded Systems*, pp. 10–19, 1999.

27. N. Kroupis *et al.* Compilation Technique for Loop Overhead Minimization, in *Proc. of Euromicro Conference on Digital System Design, Architectures, Methods and Tools (DSD)*, pp. 419–426, 2009.

28. F. Hannig and J. Teich. Design Space Exploration for Massively Parallel Processor Arrays, in *Proc. International Conference on Parallel Computing Technologies*, vol. 2127, pp. 51–65, 2001.

29. S. Boppu *et al.* Towards Symbolic Run-Time Reconfiguration in Tightly-Coupled Processor Arrays, in *Proc. International Conference on Reconfigurable Computing and FPGAs*, pp. 392–397, 2011.

30. J. Teich, A. Tanase, and F. Hannig. Symbolic Parallelization of Loop Programs for Massively Parallel Processor Arrays, in *Proc. International Conference on Application-specific Systems, Architectures and Processors*, pp. 1–9, 2013. Best Paper Award.

31. J. Teich. Invasive Algorithms and Architectures, *IT - Information Technology*, 50(5), 300–310, 2008.

Chapter 11

Some Statistical Experiments with Spatially Correlated Variation Maps

David B. Thomas

EEE Department, Imperial College London

This chapter uses Gaussian random fields as a means of modelling delay variation within chips. It is shown that for certain scales of correlation, it is possible to increase yield by spatially decorrelating the placement of components within a path.

11.1. Introduction

Peter's work on variation has always piqued my interest, particularly due to the statistical nature of the processes involved. Whenever Zhenyu or Justin brandishes a graph of variation maps [Ref. 1], I've always wondered what high-level impact we should expect — beyond the practical place-and-route implications, how does the existence of variation change what we should expect from the circuits?

This paper is a hopelessly naive attempt to translate the practical results and discussions I've seen in Peter's research discussions, into a more statistical approach. The only real result is to show that *if* spatial correlation in variance is an important factor, then logic elements should be spatially decorrelated in order to increase yield. This is a rather obvious conclusion, and I'm sure well known amongst silicon people, but as a software engineer it was fun to discover.

11.2. Properties of Uncorrelated Fields

Assume a rectangular $n \times m$ grid of components, each of which has some delay $d_{i,j}$. To start, we'll consider the delays to be independent and identically distributed (IID) Gaussian, so

$$d_{i,j} \sim N(\sigma, \mu) \tag{11.1}$$

$$\Pr[x < d_{i,j}] = \Phi((x - \mu)/\sigma), \tag{11.2}$$

where $\Phi(\cdot)$ is the standard Gaussian cumulative distribution function (CDF).

Now we'll consider any $w \times h$ sub-grid, $1 \leq w \leq n$, $1 \leq h \leq m$. Because the delays within the grid are IID, any $w \times h$ grid should be statistically identical, so the exact location doesn't matter, and we can just consider the rectangle starting at (1,1).

First let us consider the distribution of the average delay within the sub-grid:

$$\overline{D}_{w,h} = \frac{1}{wh} \sum_{x=1}^{w} \sum_{y=1}^{h} d_{i,j}. \tag{11.3}$$

This is simply the sum of wh IID Gaussians, so the average delay is also Gaussian distributed:

$$\overline{D}_{w,h} \sim N(\mu, \sigma\sqrt{wh}), \tag{11.4}$$

From a circuit point of view, what is more interesting is how bad things can get, which in terms of delay is determined by the worst delay within an area. If we define this as

$$D_{w,h}^{+} = \max_{x=1}^{w} \max_{y=1}^{h} d_{i,j}, \tag{11.5}$$

then we can describe the CDF of $D_{w,h}^{+}$ in terms of the CDF of $d_{i,j}$

$$\Pr[x < D_{w,h}^{+}] = 1 - (1 - \Pr[x < d_{i,j}])^{wh} \tag{11.6}$$

$$= 1 - [1 - \Phi((x - \mu)/\sigma)]^{wh}. \tag{11.7}$$

Similarly, the smallest delay is

$$\Pr[x < D_{w,h}^{-}] = \Phi((x - \mu)/\sigma)^{wh}. \tag{11.8}$$

11.3. Introducing Correlation

So far we have considered IID variance, so the delay of one element is unrelated to the delay of any other element. One hypothesis, which runs through Zhenyu's work, is that there is a strong spatially correlated component: if one area of silicon has higher delay, then it is likely that close components also have higher delay. To put this in a statistical framework, let us assume that the delay of each element is coupled to all immediately adjacent elements (left, up, right, down) with correlation ρ. Given this correlation structure, we need to determine some feasible covariance matrix Σ over the element delays.

The correlation matrix $A_{w,h}$ will be a positive-definite matrix, with all elements in $[-1, +1]$, and ones along the diagonal. For a $w \times h$ rectangle, the matrix will have wh rows and columns, as it needs to express every pair-wise correlation.

For 1×2 and 2×1 the covariance matrix is trivial, but for larger matrices we need to make sure that the matrix is consistent. Let us start with the matrix for a 2×2 rectangle:

$$A_{2,2} = \begin{array}{c} \\ (1,1) \\ (1,2) \\ (2,1) \\ (2,2) \end{array} \begin{array}{cccc} (1,1) & (1,2) & (2,1) & (2,2) \\ 1 & \rho & \rho & ? \\ \rho & 1 & ? & \rho \\ \rho & ? & 1 & \rho \\ ? & \rho & \rho & 1 \end{array} \qquad (11.9)$$

We have a number of entries which are not defined by our simple left-up-right-down correlation pairs, specifically the $(1,1) \leftrightarrow (2,2)$ and $(1,2) \leftrightarrow (2,1)$ correlations.

Intuitively these unknown correlations should be related to the Euclidean distance across the diagonals, but it isn't clear what exactly they should be. We could find some sort of consistent matrix, using a Kirchoff's Law-like approach, but this would have the unfortunate side-effect that correlations at the edges would be different to correlations in the centre. Assuming our original $m \times n$ rectangle was cut out of a much larger piece of silicon, it makes sense that the original local correlation process crossed the boundaries, then was sawn in half.

For this reason we'll assume that the local variation is represented by a Gaussian random field [Ref. 2]. Absent any better model for variation, an exponential correlation function will be used, which has the desirable property that correlations decrease monotonically from one. No negative correlations are ever introduced, but I'm not aware of any particular physical process which would introduce anti-correlations (though I'm sure they exist).

The continuous correlations between two points is now given as

$$\mathrm{Cor}(\vec{p}_1, \vec{p}_2) = \exp(-\lambda \times \|\vec{p}_1 - \vec{p}_2\|_2), \tag{11.10}$$

with λ acting as a scale parameter. For the elements of our grid, we'll simple discretise at integer co-ordinates to get the matrix

$$A[(x_1, y_1), (x_2, y_2)] = \mathrm{Cor}([x_1, y_1], [x_2, y_2]). \tag{11.11}$$

Armed with our new correlation structure for a grid, we can now return to the statistical properties of rectangles (woo!). The correlated elements follow a multivariate normal distribution, determined by a combination of the correlation matrix and our original parameters σ, and μ:

$$\vec{d} \sim N(\mathbf{1_{wh}}\mu, \mathbf{A}\sigma) \tag{11.12}$$

$$\Pr[\vec{x} = \vec{d}] = 2\pi^{-wh/2}|A\sigma| \exp(-1/2(\vec{x} - \vec{d})^T (A\sigma)^{-1}(\vec{x} - \vec{d})). \tag{11.13}$$

In terms of the expected mean, nothing has changed, it is still just μ.

The minimum and maximum become more complicated, as the spreading of the distribution will vary with correlation. In particular, as λ is increased, making correlations stretch longer and longer distances, the variation within the sample reduces. Figure 11.1 gives examples of random correlated fields for increasing values of λ.

This is easy to understand intuitively by considering the two extremes for λ. When $\lambda = \infty$ we find

$$\mathrm{Cor}((x_1, y_1), (x_2, y_2)) = \begin{cases} 1, & \text{if } (x_1 = x_2) \wedge (y_1 = y_2) \\ 0, & \text{otherwise}, \end{cases} \tag{11.14}$$

which is simply saying that the correlation matrix is the identity matrix, making each element independent. If all elements are independent then we are maximising the chance that one of them will

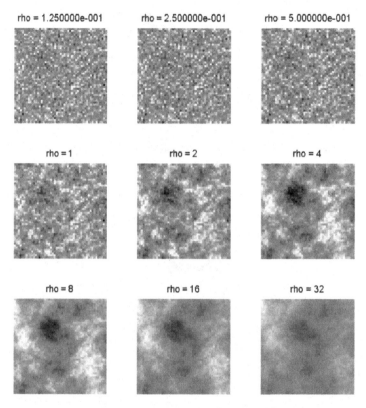

Fig. 11.1. Examples of variation as modelled by a Gaussian random field, with the same underlying random factors, but changing different correlations distances.

manage to produce a value out in the tails. By contrast, with $\lambda = 0$ we find

$$\mathrm{Cor}((x_1, y_1), (x_2, y_2)) = 1, \qquad (11.15)$$

so all elements are perfectly correlated. Perfect correlation means there is only one global source of uncertainty, so it is much less likely that a single random sample will reach into the tails of the distribution.

Figure 11.2 shows this effect visually, with each column having a different spatial correlation, and the random realisations within the column sorted by average delay. Looking at the realisations half-way up the right column, it appears as if the variance is much lower than

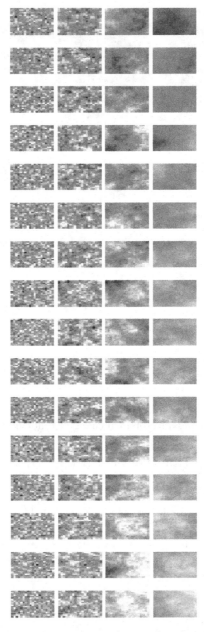

Fig. 11.2. Each column shows different realisations of a particular spatial correlation, for $\lambda = (0.125, 1, 8, 64)$. Within each column the fields are sorted by mean delay.

in the left column. However, when the entire set of realisations is considered, the overall picture becomes clearer, with some instances at the top having very low delay, and some at the bottom having very high delay.

11.3.1. *Behaviour of paths under spatial correlation*

Rectangles are useful, but a more important problem is the behaviour of paths within a circuit. We will define a path $P = p_1 \cdots p_k$ which passes through k elements as a sequence of k integer pairs, with each pair representing the location of the elements. For now we impose no restrictions on the path except that each element can only appear once.

We can define the expected delay as the sum of the delays of all components:

$$P = \sum_{i=1}^{k} d_{p_i}. \tag{11.16}$$

Under the uncorrelated model ($\lambda = 0$), then all elements are IID Gaussian, which means the variances add:

$$P \sim k\mu + N(0, \sqrt{k}\sigma), \quad \text{if } \lambda = 0. \tag{11.17}$$

At the other end, with perfect correlation ($\lambda = \infty$), all elements have the same delay, so we have

$$P \sim k\mu + kN(0, \sigma), \quad \text{if } \lambda = \infty. \tag{11.18}$$

An equivalent statement is to put it in terms of the standard deviation of path delay

$$StdDev(P) = \sqrt{k}\sigma, \quad \text{if } \lambda = 0 \text{ (uncorrelated)} \tag{11.19}$$

$$StdDev(P) = k\sigma, \quad \text{if } \lambda = \infty \text{ (perfectly correlated)}. \tag{11.20}$$

The perfectly correlated version has much higher variance for large k, which seems worse, but the truth is more subtle. The perfectly correlated version is essentially described by one random number — if that number is high, then the entire chip is pretty much useless, but there is also a 50% chance that the entire chip will be "above average", in which case every single path on the chip will run faster than average. By comparison, the uncorrelated version contains many fast and slow elements, so the different elements within a path will tend to even out the delay. However, across many paths in a chip, the likelihood that any single path is slow grows quite quickly.

To examine $0 < \lambda < \infty$, it is necessary to consider what paths to check, as the spatial nature of the paths will affect the delay distribution for the paths. Here we'll consider an abstract design consisting of four elements connected in a chain, and our yield metric will be based on tolerable deviation from the mean. Based on static analysis, with no knowledge of the actual chip variation, the tools could assume that the standard deviation of each path will simply be $\sqrt{k}\sigma$, so in this case 2σ. Let us assume the tools work to a slack s, and any given four element circuit will work as long as $P < 4\mu + s$.

The target design contains multiple paths, and a given chip only works if all paths meet the slack target. Because each chip will have a different delay distribution, we'll define a yield metric $Y(s)$. The yield estimates the number of chips which will function correctly when designed for a particular slack.

In this simple study, there are only four elements in each path, and it is natural to pack them together into a square. Figure 11.3(a) shows this layout. Another common layout is to arrange them into columns (e.g. carry-chains), shown in (b). While these arrangements should be equivalent in an uncorrelated or perfectly correlated field, in spatial fields there is a potential difference: if one of the elements in a 2×2 block happens to get a large delay, it is also likely that the other elements will have high delays, so the chance of the entire path exceeding slack is much higher. In contrast, the column has a greater chance of staying within slack, as if the top of the column has a high delay there is still some hope that the bottom will be far enough away to be decorrelated, and have a much smaller delay.

The idea that space between elements increases the likelihood of any given path working suggests that other arrangements might be even more efficient. For example, if we were to take pairs of columns and swap every other element, then the zig-zag arrangment shown in Fig. 11.3(c) arises. Although superficially similar, now all elements in

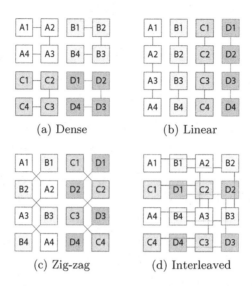

(a) Dense   (b) Linear

(c) Zig-zag   (d) Interleaved

Fig. 11.3. Layouts for paths of four elements, contained within a 4×4 grid.

a path are at least a distance $\sqrt{2}$ from each other, which means the maximum pair-wise correlation drops from $\exp(-\lambda)$ to $\exp(-\lambda\sqrt{2})$.

Going further, it is possible to interleave the paths as shown in (d), though at the cost of more routing. Now the maximum correlation is $\exp(-\lambda 2)$, a great improvement over the original.

Given these four arrangements, we can calculate their yield for different λ and sizes empirically, by generating lots of different variation maps. The maximum path delay within any variation map will determine the minimum allowable slack, and counting the number of random instances with slack less than s will provide an estimate of $Y(s)$.

Figure 11.4 shows the changing yield for a device with 4×4 elements, with each device used to hold four paths. The gently curving yellow line across the middle is the completely correlated case, which doesn't reach high yields even for very large slack. The upper line is the completely uncorrelated case, which provides the best possible yield. Both these cases are independent of the placement. The remaining lines show the effect of the different placements when $\lambda = 1$. Each placement provides a small increment

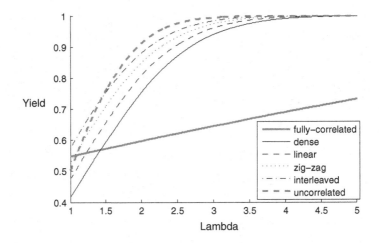

Fig. 11.4.   Yield vs. λ for a 4×4 device.

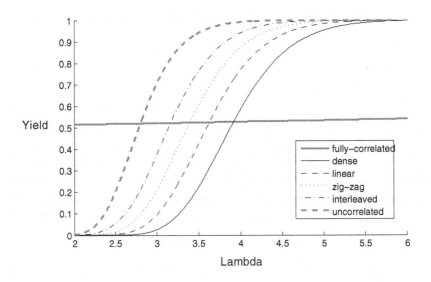

Fig. 11.5.   Yield vs. λ for a 32×32 device.

in yield, for example, at a slack of 2, moving from dense placement
to interleaved placement provides about a 10% increase in yield.

Figure 11.5 performs the same experiment, but now with a 32×32
grid. Because there are now so many more independent paths to test,
the chance of any one of them requiring large slack is much higher, so

the benefits of removing spatial correlation are larger. At a slack of 3.5 the yield increases by 50% if the dense placement is replaced with the interleaved. Even the much less drastic conversion from linear to zig-zag has a significant effect, with yield rising by just over 10%.

11.4. Conclusion

We assumed that delay due to variation could be modelled with a spatially correlated Gaussian random field, and then looked at the yield of chips containing many identical paths. Superficially, the maps look like those extracted from the DE0 boards by Peter's research group [Ref. 1], but it would be interesting to test the hypothesis more rigorously. Attempting to characterise the scale of variation would also be interesting, as the ability to deal with spatial correlation depends a lot on how far it reaches. There is also an interaction with the reliability and variation of the interconnect [Ref. 3], so modelling both at the same time might suggest that spatial decorrelation is a bad idea.

My original goal in writing this chapter was to look at the statistical properties of self-repair mechanisms in the context of spatially correlated variation, using some of the approaches used for wear-levelling [Ref. 4]. However, time did not permit, so this is left for future discussion with Peter's research group

References

1. Z. Guan *et al.* A Two-stage Variation-aware Placement Method for FPGAs Exploiting Variation Maps Classification, in *Proc. International Conference on Field Programmable Logic and Applications*, pp. 519–522, 2012.
2. P. Abrahamsen. A Review of Gaussian Random Fields and Correlation Functions. [Online] Available at: http://www.math.ntnu.no/ omre/TMA4250/ V2007/abrahamsen2.ps. [Accessed 23 April 2014].
3. N. Campregher *et al.* Yield Modelling and Yield Enhancement for FPGAs Using Fault Tolerance Schemes, in *Proc. International Conference on Field Programmable Logic and Applications*, pp. 409–414, 2005.
4. E. Stott and P.Y.K. Cheung. Improving FPGA Reliability with Wear-levelling, in *Proc. International Conference on Field Programmable Logic and Applications*, pp. 323–328, 2011.

Chapter 12

On-Chip FPGA Debugging and Validation: From Academia to Industry, and Back Again

Steve Wilton

Department of Electrical and Computer Engineering
University of British Columbia

In recognition of Peter Cheung's 60th birthday and his extensive experience in digital system design research and education, this chapter discusses recent developments in on-chip debugging with a vision towards the future. We focus primarily on techniques to increase observability but also touch upon other aspects of the problem. We end with a discussion of the potential of extending debugging techniques to the world of high-level synthesis generated designs.

12.1. Introduction

Unprecedented advances in integrated circuit fabrication technology has changed our way of life. Mobile computing, the high speed internet, and computing equipment that analyses and manipulates information has changed the way we do business, relax, and interact with each other across the planet. The impact of increased computing power is like no other revolution our society has ever seen. Despite technology challenges, these advances show no sign of slowing down; in fact, the dramatically increased rate of communication has actually accelerated the design of new technology that underpins future advances.

A key challenge with these new technologies, however, is keeping the cost of exploiting the technology affordable. As on-chip integration increases, the complexity of the underlying technology has also increased, leading to dramatically increased costs and financial risk. Today, very few companies are capable of creating high-end integrated circuits. For everyone else, field-programmable gate arrays (FPGAs) have become the implementation medium of choice for many digital circuits. FPGAs can be configured to implement any digital circuit, allowing designers to immediately test designs without the cost, risk and delay of producing a VLSI implementation. They provide companies with large-scale integration without requiring access to a state-of-the-art chip fabrication plant. The improvement of FPGA technology, the associated CAD algorithms, and ways in which FPGA technology can be used, has formed the backdrop for much of Peter Cheung's research over the past several decades.

One of the key challenges in designing large digital systems is verifying that the designs are correct (*verification* and *validation*), and finding the root cause of design errors when they are observed (*debugging*). A recent study from Mentor Graphics showed that half of all design effort was used for functional verification [Ref. 1], and that the situation is getting worse — designer productivity doubles only every 39 months [Ref. 1], while silicon density doubles every 18 months due to Moore's Law. This has led many researchers around the world to investigate techniques to accelerate the verification, validation, and debugging of digital systems.

In this paper, we focus on one aspect of this puzzle: debugging. As we will describe, a primary challenge in debugging large digital systems is lack of *visibility* into the internal state of a system. This has motivated much work, and we will discuss some of it here. To make the discussion concrete, we will use an example case study that traces a particular technology development from an academic lab, to an industry startup, through a successful acquisition, and then back to the industrial lab. Although the focus will be on our own research, this type of research is very much related to the overall problems being studied by Peter Cheung; we hope that by reading this paper, the audience can better understand some of the challenges in this field — challenges that Peter works to address every day.

12.2. The Need for On-Chip Debugging

When digital systems were small, it was possible to test and debug these circuits entirely in simulation. Today, this is rarely possible [Ref. 2]. Using today's technology, the task of booting Linux on a SoC would take roughly 2,000 years to simulate. To put this in context, when the Romans were in London, if they had started a simulation in Peter's lab, it would be completing "any day now" (this is assuming the Romans had computing technology equivalent to today's, which seems unlikely). If the simulation showed an error, and had to be re-run, we would need to wait another 2,000 years to get the results, significantly delaying the graduation of the students involved. Yet, it is easy to imagine that there are many bugs that can not be "activated" without booting an operating system (at least). Clearly, simulation (even when accelerated) is inadequate to thoroughly exercise any large digital system.

A second challenge with simulation is that it is difficult, if not impossible, to test the system *in-situ*. Many bugs will only be apparent when the digital system receives real-world stimulus. Trying to recreate the stimulus in a testbench is often difficult, and suffers from the problem that it is the *unexpected* stimulus that often causes errors, and those types of inputs are unlikely to be encoded in a testbench. Secondly, most digital systems operate in an ecosystem with either other chips or other embedded intellectual property (IP) blocks; bugs often occur because designers fail to understand (exactly) the interface requirements of these blocks (often undocumented) or the peculiarities of how the blocks should be used (again, often undocumented).

Finally, any interaction with embedded software or firmware is a common source of buggy behaviour, yet those interactions are often difficult or impossible to adequately cover without running a real system.

For all these reasons, the only way to completely test a systems, and the only way to find many bugs, is using an actual working chip.

12.3. Key Challenge: On-Chip Visibility

A key challenge when debugging a real working system is *observability*. Moore's Law continues to provide more transistors (despite many obstacles); however, the rate of increase of I/O connections on a chip is not increasing as fast. The situation may get worse with the emerging 2.5-D and 3-D packaging technologies, simply because there is more logic inside one package.

In simulation, visibility is not an issue, since the user can "probe" any signal he or she wants to observe in the software model of the system. This is a vital step in the debugging of a circuit; observing internal signals is the best way for an engineer to "narrow down" their understanding of the potential state a circuit may be in, in order to help deduce the cause of unexpected behaviour (indeed, this is how we teach our undergraduates to debug; look at important signals to help narrow down the cause of an error).

In a hardware system, unless a signal is connected to an external pin, it is impossible to observe (chip probing equipment does exist, but this is expensive, error-prone, and is often limited to observing only the top few layers of metal). In our undergraduate classes, we teach students how to connect internal signals to pins during debugging (and then recompiling their design), but this is time-consuming, and limits debug productivity.

12.4. Case Study: From Academia to Industry

To make our discussion concrete, in this section we trace the development of a particular technology (or family of technologies) that address this problem, starting from its inception in academia, to its exploitation in the startup Veridae Systems.

12.4.1. *Trace buffers and monitoring circuits*

Observability can be enhanced using embedded *trace buffers*, as shown in Fig. 12.1. The idea is to connect signals that are deemed "important" to one or more embedded memories, and record the values of those signals during normal operation [Refs. 3 and 4].

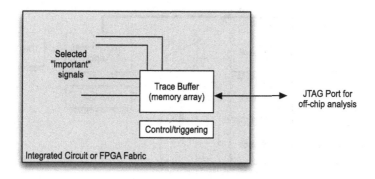

Fig. 12.1.   Integrated circuit instrumented with a trace buffer.

When an error is observed, the system can be halted (halting a system precisely is difficult; addressing this "skidding" problem is an interesting research opportunity), and the history of these signals can be read out. Tools such as SignalTap II and ChipScope are produced by FPGA vendors to automate the inclusion of this instrumentation and other forms of instrumentation have been described [Ref. 5].

There are a number of challenges with this approach. First, only a limited amount of data can be recorded in a trace buffer. There has been work on both lossless and lossy compression for trace buffer data, which increases the amount of information stored [Ref. 6].

A second challenge is the need to determine, *a priori*, which signals are "important" so that these signals can be hardwired to the trace buffers. This *signal section* problem has been well-studied but remains difficult [Refs. 7–10].

12.4.2. *Academic efforts: Embedded logic analyzers*

To address some of these problems we proposed the use of *programmable logic cores* as embedded logic analysers, as shown in Fig. 12.2. The idea is to incorporate a programmable logic (FPGA) region within a fixed-function chip, such as an application-specific integrated circuit (ASIC). This programmable logic region can be configured to implement any digital circuit at run-time by setting the values of configuration bits, just as can be done in an FPGA. At

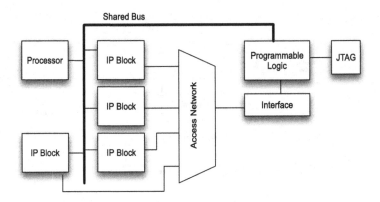

Fig. 12.2.   Integrated circuit instrumented with programmable logic fabric.

run-time, when an engineer is searching for the cause of unexpected behaviour, he or she can configure this region (perhaps, for example, creating complicated gating functions) that record trace buffer data only when a specific pattern on a number of other signals (perhaps over time) is observed. The programmable logic region could also be used to compress data, count the number of predetermined events that occur, or implement assertions that might be helpful during debug [Ref. 11].

Furthermore, it is conceivable that, under certain circumstances, the programmable logic core could be used to *over-write* certain values. This provides a measure of *controllability* to the debug process. One example where this might be useful during debug is to suppress an error signal that is activated, possibly allowing the system to "limp along" after an error occurs.

This technique still requires selecting signals to hardwire to the programmable logic core. In our case, we proposed using a *concentrator* network which can be configured, at run-time, to efficiently connect a larger number of signals (on the order of 1,000s) to the smaller number of embedded programmable logic core pins [Ref. 12]. Since this concentrator can be configured at run-time, it is possible to change the set of signals used for each debug scenario.

There are several challenges with this approach. First, the programmable logic fabric is large. Research has suggested that programmable logic can be 35 × larger than the equivalent ASIC circuitry [Ref. 13], so this limits the complexity of the debug logic that can be implemented in a core. However, we were able to show that by limiting the overhead to 10% (this was deemed acceptable by many of our industrial contacts, since this extra logic does not need to yield in a production environment), we could implement a large number of debug scenarios.

12.4.3. *Industrial efforts: Shifting the technology*

In 2011, we decided to commercialise this approach, and created a Vancouver-based company called Veridae Systems. Veridae started out at the University of British Columbia, but quickly moved to an off-campus office. Initially, there were six programmers, hired using our own seed money and various government grants and related sources.

As we began to commercialise the technology, it became clear that the technology developed at the university was not suitable for our purposes. In particular, we quickly determined that a general programmable logic core was far more flexible than was necessary. Although we wanted to support a variety of debug scenarios, most of these debug scenarios were very similar, and so much of the programmable logic could be replaced by much more efficient hard logic. In particular, we created very complex compression circuitry out of hard logic (in typical cases, our compression circuitry made it possible to gather data for *seconds* of run-time by filtering out uninteresting events on a bus, for example). We also found that the full flexibility afforded by the programmable logic core was not needed for common matching applications; we thus replaced this logic with fixed hand-optimised flexible mapping circuitry. The concentrator network also evolved as we commercialised the technology. As a result, the eventual technology that was commercialised looked very different from the academic work that had led to the company in the first place [Ref. 14].

The other shift in the technology was the focus on usability. In our academic work, usability was not a key concern, however, within the company, one of our primary values to our customers was the ability to easily and quickly insert debug instrumentation; thus, creating a intuitive yet flexible interface became important. A very significant part of our development was aimed towards creating a GUI infrastructure (based on Python and QT).

Another difference between the industrial and academic effort is that in the academic world, it was sufficient if our tools worked on just a "handful" of user circuits. In the industrial world, our tool had to work on *all* user circuits, whether they be written in VHDL, Verilog, or System Verilog. We quickly came to appreciate the semantic differences (differences in the way code can be written) available to designers in all three languages. We used software from Verific to help in this process, however, we found that a significant amount of effort was spent ensuring our tool handled all corner cases of all three languages correctly.

12.4.4. *Industrial efforts: Shifting the priorities*

A second shift occurred as we commercialised our technology. Although we had initially focused on ASIC designers, it quickly became clear that (a) there are not many ASIC designers today, since more and more design starts are shifting to FPGAs, and (b) the design cycles for ASICs are much longer than for FPGAs, meaning it is hard for a startup to have the right timing to seamlessly fit into a customer's design schedule. Thus, a key strategic decision was to put significant effort into an FPGA product.

A second strategic decision was the realization that there are many ASIC designers that use *FPGA prototyping* to test and debug their designs. Recently, Intel has described how their Atom processor was prototyped on a single FPGA running at 50 Mhz [Ref. 15], and how their i7 Nehalem processor was prototyped using five FPGAs running at 520 kHz [Ref. 16]. These sorts of prototyping systems have become essential as designers create larger and more complex designs [Refs. 17 and 18]. It became apparent that it was during FPGA prototyping that many bugs were uncovered, so the company also put resources into addressing that market. A key challenge when addressing the FPGA prototyping market was to handle asynchronous interactions between different devices, and to later reconstruct signals based on measured time references.

12.5. Back to Academia: Future Directions

In 2011, Veridae Systems was sold to Tektronix. Our experience in industry motivated a significant amount of new academic work, including more efficient incremental trace techniques, overlay networks, and better signal selection methods. We also considered a related problem; coverage monitoring. In this section, we highlight some of this future work, some of which has begun, and some of which is still in its infancy. For readers not working in this field, we hope this section can provide an appreciation of the breadth and excitement of some of the work remaining to be done in this area. Perhaps this section may even provide some motivation for Peter to start addressing some of these challenges in the future.

12.5.1. *Coverage monitoring*

In the software design world, companies regularly use "code coverage" to determine how effectively a set of tests "cover" a design. It is not immediately clear how to translate this to a hardware design environment. Coverage techniques such as *statement coverage, branch coverage, condition coverage, path coverage, functional coverage, mutation coverage* and *tag coverage* have all been proposed.

Each of these coverage metrics can be easily measured using simulation techniques; however, for the same reasons as described in the previous section, effective coverage measurements can only be done using real running hardware. Hardware validation is accepted as the only method to thoroughly exercise a design. However, gathering these coverage numbers through validation requires instrumentation possibly with significant overhead.

As a specific example, consider *statement coverage*. An interesting question would be *when I boot Linux on my SoC, what statement coverage do I achieve?* The answer to this question would give an

indication of whether booting Linux is a "useful" test, and what
other tests need to be performed before a design is deemed correct
(or correct enough to ship). However, from the above, measuring this
in simulation would take thousands of years. A naïve approach would
entail adding instrumentation for every instruction that would trigger
if this instruction has been "exercised". We have shown [Refs. 19
and 20] that the overhead can be reduced by instrumenting at the
basic block level, and further reduced using compiler techniques
such as those by Agrawal [Ref. 21]. However, the overhead is still
unreasonable, especially if it is not economically feasible to create
a new spin to remove the instrumentation before shipping. Clearly
more research is needed here.

Another approach is to employ an FPGA prototype, and measure the coverage using the FPGA prototype. One challenge here
(especially when it comes to coverage metrics such as path coverage)
is that the FPGA implementation of the user circuit may not
correspond exactly (gate to gate) to the original ASIC implementation. When mapping the design to an FPGA prototyping platform,
different optimisations are performed to ensure the FPGA resources
are used efficiently. Nonetheless, it is possible to instrument the
FPGA implementation to gather information regarding the coverage,
as would be seen if this test suite was run on the original ASIC.

A final opportunity is the ability to unify the debug infrastructure,
such as that described earlier in this paper, and the instrumentation
required to measure coverage. Indeed, measuring coverage is about
visibility, so techniques that can be used to enhance visibility during
debug can also be used to measure the coverage of a set of tests. This
is interesting work that has received little attention.

12.5.2. *Future directions: Debugging from high-level synthesis*

Decreased electronic product life cycles necessitate shorter times
from product conception to market introduction. It is clear that
"time to market" is becoming increasingly important. To address
this, FPGA vendors have recently invested in high-level synthesis

(HLS) technologies that automatically transform a software program into a hardware circuit (e.g. Altera has OpenCL support and Xilinx supports AutoESL C-based design flows). Raising the abstraction level for design-entry to support software-like languages has two advantages. First, it has been reported that there are 10 × more software developers than hardware developers [Ref. 22]; increasing accessibility to this large labour pool allows more companies to leverage hardware's dramatically higher data throughput and energy efficiency benefits. Second, a higher abstraction level increases the productivity of each designer, just as the move from gate-level design to register-transfer level design did in the 1980s. This new technology is disruptive and could dramatically change the computing landscape [Ref. 23].

However, this abstraction technology can only be effective if it is part of an ecosystem in which designers using software-like methodologies can visualise and analyse their design's behaviour. Such an ecosystem requires that software engineers can create correct and efficient hardware structures, without requiring detailed knowledge of digital hardware design. Hardware design is characterised by cycle-by-cycle operations, low-level optimisation, and a dataflow-oriented view of a computing problem. However, software engineers typically view designs as interacting sequential processes, ignoring their cycle-by-cycle behaviour, low-level mapping and operations. An ecosystem that spans these two very different abstraction layers is essential to the accelerated time-to-market promised by HLS-based technologies.

12.5.2.1. *Motivation*

We believe there is a critical need to investigate and develop key technologies for this ecosystem. Using today's tools, it is possible for a software designer to use HLS tools to create hardware running on an FPGA programmable platform by writing only software. However, debugging is still challenging. Today, a common test and debug methodology is to port, compile, and run the code directly on a workstation to verify the functionality and help find bugs before compiling to hardware. Although appropriate for initial verification, this emulation approach has two limitations. First, the software emulation will run much slower than the target hardware (typically 20 to 200 times slower) limiting the thoroughness of tests that can be performed. Second, this will not uncover problems related to interactions with the environment or with other modules in the system; experience from our company indicates this is where most bugs occur. Thus, it is necessary to test the design *in situ* by executing the synthesised hardware on the target FPGA.

Today, in order to perform this hardware verification, a designer would compile the software using a HLS tool (e.g. LegUp [Ref. 24]) to a hardware specification (more precisely, a register transfer level (RTL) design). This specification can then be mapped to hardware using standard tools. Debugging packages such as the one described earlier in this paper can then be used to provide visibility into the design to help the engineer understand the operation of the hardware to narrow down the cause of a suspected bug.

The challenge with this approach is that these tools provide visibility that has meaning only in the context of the generated RTL hardware. A software designer typically would not have an understanding of the underlying hardware; in fact, this is the primary reason that HLS methodologies are able to deliver high design productivity. A software designer's perspective centres on sequentially executed statements, written in a high-level language, to realise computations and control flow — there is no notion of a clock. On the other hand, the key advantages of hardware arise from the use of interconnected dataflow components operating in parallel across multiple clock cycles. Further, HLS reschedules operations across clock boundaries, creating a disconnect between the software- and hardware-views of the state of a system. A C-level instruction is not mapped to the hardware as a unit, but rather is optimised along with other instructions to create composite hardware units that execute within a single clock cycle. We argue that this mismatch between how a software designer views a design, and how current on-chip debugging tools present results, is the most important challenge facing HLS methodologies today. In the long term, this will limit the ability of a large class of designers (software designers) to obtain the performance and power advantages of FPGA programmable platforms.

12.5.2.2. *Key technical challenges*

The key technical challenge is to create a bridge between a software view of a design and the synthesised hardware. This can be addressed by developing and employing two complementary technologies: instrumentation and transforms.

As described earlier in this paper, effective debug and optimisation can best be achieved through instrumentation, that is, by adding small amounts of logic to a circuit/program to provide visibility and controllability to the design. The tool we described inserts instrumentation into the RTL (hardware) circuit; as a result, the instruments gather and control signals at the hardware level, creating the disconnect. It is also possible to instrument at other levels of abstraction — in particular, if we instrument the original software, we would immediately gather and control signals that the software programmer understands, however, this would significantly constrain the optimisation opportunities available to the HLS compiler, leading to slower, larger, and more power hungry circuits. Likely, the best solution combines instrumentation at various levels of abstractions, balancing the overhead versus of observability and controllability.

Another approach to bridge this disconnect is to develop and use transforms that relate structures and time references in the hardware circuit to those in the software design. Work by Hemmert [Ref. 25] provides the groundwork for creating such transforms. However, his techniques were not designed or evaluated in the context of a commercial HLS tool that performs more than 50 optimisations before the hardware is generated (it is these optimisations that create the disconnect we wish to address). Nonetheless, Hemmert's work provides optimism that such transforms are possible and can be effective.

12.5.3. *Integration with run-time binding*

In the past few years, Peter Cheung has been doing important work in the area of run-time binding [Refs. 26 and 27]. Through run-time binding it is possible to tolerate, and even take advantage of, manufacturing differences between individual devices (faults or simply parametric differences).

An exciting research direction would be to unify on-chip debugging and visibility-enhancement techniques and Peter's run-time binding work. Run-time binding requires knowledge of the particulars of the device being used, and it is conceivable that this sort of information could be gathered using an infrastructure like those used for debug in this paper. Rather than trace buffers, it may be possible to integrate sensors such as those developed by Peter and his group [Refs. 28–31]. In the long-term, this may lead to systems that not only detect errors, but are more reliable [Ref. 32], and eventually those that *self-heal*.

12.6. Conclusions

The world is a different place than it was even a few decades ago. We are now all wirelessly connected to each other, around the world, and this has had a dramatic impact on not only how we work, but how we relate to each other and live our lives. This transformation has been enabled by technology. Throughout (the first part of) his career, Peter Cheung has been central to this revolution. He has made extremely important contributions that have helped take us into this new world.

As an example of the types of challenges that we need to address to continue this revolution, this chapter has talked about techniques to accelerate the debug of integrated circuits. We focused on a specific case study, and traced the development of a particular technology from academia to industry, and then showed how the circle was completed by inspiring new research in the academic realm. We hope that some of the ideas in this paper might be inspiring for others, both within Peter's research group and beyond.

References

1. H. Foster. Challenges of Design and Verification in the SoC Era, Report, Oct. 2011.
2. A. Nahir, A. Ziv, and S. Panda. Optimizing Test-generation to the Execution Platform, in *Proc. Asia and South Pacific Design Automation Conference*, pp. 304–309, 2012.
3. M. Abramovici *et al.* A Reconfigurable Design-for-Debug Infrastructure for SoCs, in *Proc. Design Automation Conference*, pp. 7–12, 2006.
4. H. Ko, A. Kinsman, and N. Nicolici. Design-for-Debug Architecture for Distributed Embedded Logic Analysis, *IEEE Transactions on Very-Large Scale Integration Systems*, 19(8), 1380–1393, 2011.
5. E. Matthews, L. Shannon, and A. Fedorova. A Configurable Framework for Investigating Workload Execution, in *Proc. International Conference on Field-Programmable Technology*, pp. 409–412, 2010.
6. E. Anis and N. Nicolici. Low Cost Debug Architecture Using Lossy Compression for Silicon Debug, in *Proc. Design Automation & Test in Europe Conference & Exhibition*, pp. 1–6, 2007.
7. H.F. Ko and N. Nicolici. Algorithms for State Restoration and Trace-Signal Selection for Data Acquisition in Silicon Debug, *IEEE Transactions on Computer-Aided Design of Circuits and Systems*, 28(2), 285–297, 2009.
8. X. Liu and Q. Xu. Trace Signal Selection for Visibility Enhancement in Post-Silicon Validation, in *Proc. Design Automation & Test in Europe Conference & Exhibition*, pp. 1338–1343, 2009.
9. E. Hung and S. Wilton. On Evaluating Signal Selection Algorithms for Post-Silicon Debug, in *Proc. International Symposium on Quality Electronic Design*, 2011.
10. S. Wilton, B. Quinton, and E. Hung. Rapid RTL-based Signal Ranking for FPGA Prototyping, in *Proc. International Conference on Field-Programmable Technology*, pp. 1–7, 2012.
11. M. Boulé and Z. Zilic. *Generating Hardware Assertion Checkers: for Hardware Verification, Emulation, Post-Fabrication Debugging and On-Line Monitoring*, Springer, New York, NY, USA, 2008).
12. B. Quinton and S. Wilton. Concentrator Access Networks for Programmable Logic Cores on SoCs, in *Proc. International Symposium on Circuits and Systems*, pp. 45–48, 2005.
13. I. Kuon and J. Rose. Measuring the Gap between FPGAs and ASICs, *IEEE Transactions on Computer-Aided Design of Integrated Circuits and Systems*, 62(2), 203–215, 2007.
14. B. Quinton, A. Hughes, and S. Wilton. Post-Silicon Debug of Complex Multi Clock and Power Domain ICs, in *Proc. International Workshop on Silicon Debug and Diagnosis*, 2010.
15. P. Wang *et al.* Intel Atom Processor Core Made FPGA-Synthesizable, in *Proc. International Symposium on Field-Programmable Gate Arrays*, pp. 209–218, 2009.

16. G. Schelle *et al.* Intel Nehalem Processor Core Made FPGA Synthesizable, in *Proc. International Symposium on Field-Programmable Gate Arrays*, pp. 3–12, 2010.

17. S. Mitra, S. Seshia, and N. Nicolici. Post-Silicon Validation Opportunities, Challenges, and Recent Advances, in *Proc. Design Automation Conference*, pp. 12–17, 2010.

18. B. Heaney. Keynote Address: Designing a 22 nm Intel Architecture Multi-CPU and GPU, in *Proc. Design Automation Conference*, 2012.

19. K. Balston *et al.* Post-Silicon Code Coverage for Multiprocessor System-on-Chip Designs, *IEEE Transactions on Computers*, 62(2), 242–246, 2013.

20. M. Karimibiuki *et al.* Post-silicon Code Coverage Evaluation with Reduced Area Overhead for Functional Verification of SoC, in *Proc. High Level Design Validation and Test Workshop*, pp. 92–97, 2011.

21. H. Agrawal. Dominators, Super Blocks, and Program Coverage, in *Proc. Symposium on Principles of Programming Languages*, pp. 25–34, 1994.

22. U.S. Bureau of Labour Statistics. *Occupational Outlook Handbook*, Report, 2012.

23. Q. Liu *et al.* Compiling C-like Languages to FPGA Hardware: Some Novel Approaches Targeting Data Memory Organization, *The Computer Journal*, 54(1), 1–10, 2011.

24. A. Canis *et al.* LegUp: High-level Synthesis for FPGA-based Processor/accelerator Systems, in *Proc. International Symposium on Field Programmable Gate Arrays*, pp. 33–36, 2011.

25. K. Hemmert *et al.* Source Level Debugger for the Sea Cucumber Synthesizing Compiler, in *Proc. International Symposium on Field-Programmable Custom Computing Machines*, pp. 228–237, 2003.

26. Z. Guan *et al.* A Two-Stage Variation Aware Placement Method for FPGAs Exploiting Variation Maps Classification, in *Proc. International Conference on FieldProgrammable Logic and Applications*, pp. 519–522, 2012.

27. P.Y.K. Cheung. Process Variability and Degradation: New Frontier for Reconfigurable, *Reconfigurable Computing: Architectures, Tools and Applications*, 5992, 2, 2010.

28. J. Levine *et al.* Online Measurement of Timing in Circuits: for Health Monitoring and Dynamic Voltage and Frequency Scaling, in *Proc. International Symposium on Field-Programmable Custom Computing Machines*, pp. 109–116, 2012.

29. J. Levine *et al.* Health Monitoring of Live Circuits in FPGAs Based on Time-Delay Measurement, in *Proc. International Symposium on Field-Programmable Gate Arrays*, pp. 284–284, 2011.

30. J.S.J. Wong, P. Sedcole, and P.Y.K. Cheung. Self-Measurement of Combinational Circuit Delays in FPGAs, *ACM Transactions on Reconfigurable Technology and Systems*, 2(2), 10:1–10:22, 2009.

31. J.S J. Wong, P.Y.K. Cheung, and P. Sedcole. Combating Process Variaton on FPGAs with a Precise Delay At-speed Test Measurement Method, in *Proc. International Conference on Field-Programmable Logic and Applications*, pp. 703–704, 2008.

32. E. Stott and P.Y.K. Cheung. Improving FPGA Reliability with Wear-Leveling, *in Proc. International Conference on Field-Programmable Logic and Applications*, pp. 323–328, 2011.

Chapter 13

Enabling Survival Instincts in Electronic Systems: An Energy Perspective

Alex Yakovlev

School of Electrical and Electronic Engineering,
Newcastle University

The writing of this chapter has been inspired by the motivating ideas of incorporating self-awareness into systems that have been studied by Professor Cheung in connection to dealing with variability and ageing in nano-scale electronics. We attempt here to exploit the opportunities for making systems self-aware, and taking it further, see them in a biological perspective of survival under harsh operating conditions. Survivability is developed here in the context of the availability of energy and power, where the notion of power-modulation will navigate us towards the incorporation into system design of the mechanisms analogous to instincts in human brain. These mechanisms are considered here through a set of novel techniques for reference free sensing and elastic memory for data retention. This is only a beginning in the exploration of system design for survival, and many other developments such as design of self-aware communication fabric are further on the way.

13.1. Introduction

Complex information and communication systems have been studied for a long time. Many approaches and methodologies for their modelling, analysis and design exist to date. Amongst the properties of interest in those studies a prominent place is occupied by the property of systems to stay alive and function in spite of harsh environmental conditions that may surround them. Typically such conditions are assumed to generate higher rates of errors, such as those that are for example caused by radiation. They are considered mostly in the scope of information processing, and to a lesser extent in the domain of resource availability (for example, the availability of energy, the mother of all resources). While the system may remain fully functional under the nominal conditions of energy supply, its behaviour may be highly unpredictable when the energy flow to the system is impaired for one reason or another. Design of systems with varying power modes is a rapidly emerging area of research, and it comes from many different directions; for example, intelligent autonomous systems, systems with energy harvesting, green computing etc. Much of this research is about systems that are sufficiently complex that even their most energy-frugal mode of action still requires a certain stable level of energy flow. What about systems that have to 'live on the poverty line', the conditions in which power levels drop to zero and systems that have to self-recover upon the arrival of the 'first beam of sunlight'?

In this paper we shall look at the first glimpses of, perhaps still naive, approaches to building electronic computer systems whose power sources can be defined in a wide band of modes. Such systems will effectively need survival instincts as part of their intrinsic characteristics. An important element of this new design discipline is a close link between the design methods required for power conditioning and those necessary for computational blocks, as the latter form the load in the overall power chain. This proximity and even interplay of energy and information flows, and the associated holistic nature of system development activities, is what drives us towards a new type of co-design, which involves new methods for modelling, simulation, synthesis and hardware and software implementation. This chapter will address a number of paradigms for such designs, such as power-modulated computing and elastic system design. It will present examples of problems formulated and solutions obtained in the context of research on the new generation of systems with higher self-awareness for survivability. A prominent place in this exploration is taken by what we call reference-free sensing, which allows the system to check its power conditions without relying on external references in voltage or clock.

On-chip sensing is generally a very important area of research in modern times due to the high variability of devices produced in nanometer technologies. Before any piece of fabricated silicon is put into action, it has to be measured and tuned to help its performance best meet its individual characteristics. Ageing is another factor that requires adaptation of functional settings, voltage and frequency scaling, throughout the lifetime of the system. This has been realised by Professor Peter Cheung and his co-workers at Imperial College who investigate methods for health monitoring of chips, exploring their individual character and looking for ways of run-time performance optimisation (e.g. [Ref. 1]). In many respects the various built-in self-awareness facilities for adaptation to variations and ageing are similar to those for survival. This interesting relationship and long term professional friendship with Professor Cheung have inspired the author in writing this chapter for such a wonderful occasion!

Before we start our journey into the subject of this work, it would be pertinent to bring two important quotations:

*"The very essence of an instinct is that it is followed **independently of reason**."* 1871: C. Darwin *Descent of Man* I. iii. 100.

*"The operation of instinct is **more sure and simple than that of reason**."* 1781: E. Gibbon *Decline & Fall* (1869) II. xxvi. 10.

We bring these quotations with one purpose: for our study of certain basic functionalities in electronic systems that are retained in the conditions of austerity, we need an analogy with biology. The biological world is the realm where survival is a key property of organisms, whether it concerns each organism individually or organisms as a species. As we postulated above, instincts are seen as something which is inherent to survival. So is the importance of these quotations — they define the place and role of instinct along and in comparison with reason, something that is regarded as the highest form of biological activity. Armed with this analogy, we will start looking at the ways that electronic systems can be built where their 'reason' parts operate along with their 'instinct' parts. The outline of topics discussed in this chapter is as follows:

- Bio-inspiration: survival instincts in real life.
- 'Survival instincts' in ICT systems.

- Energy-power modulation and layers of functionality.
- Mechanisms in energy and data processing:
  - Reference-free sensing,
  - Elastic memory for data retention,
  - Elastic power supply for survival.
- Future developments.

13.2. Survival and Instincts in Real Life

So, what are survival and instinct in general terms? Among the many definitions of survival and instinct that can be found in the OED, perhaps the following serve our needs best: "Survival: the continuing to live after some event; remaining alive, living on". "Instinct: (a) an innate propensity in organized beings (esp. in the lower animals), varying with the species, and manifesting itself in acts which appear to be rational, but are performed without conscious design or intentional adaptation of means to ends. Also, the faculty supposed to be involved in this operation (formerly often regarded as a kind of intuitive knowledge). (b) Any faculty acting like animal instinct; intuition; unconscious dexterity or skill".

If we were looking at instincts from a biological or even psychological perspective, we would have distinguished between instinct and intuition. In our present analysis, we will also do that, and see intuition as, perhaps, the highest form of instinct that is close to reasoning. It is akin to prediction in information systems, which often connects higher forms such as reasoning with sensory-signalling forms. In our analysis we will not go to the level of intuition analogy, but rather stay at the level of basic instincts. What's more we will mostly approach instincts from the perspective of energy in the system, and see how energy or power levels determine the role of instincts, particularly focussing on their manifestation under the low energy conditions.

To get a better sense of how instincts may reveal themselves both structurally and behaviourally, we illustrate them in the following way. Firstly, we bring an example of a 'case study' which shows the energetic aspect of instincts quite vividly. A few years ago, the world heard a story about a French cave explorer Jean-Luc Josuat, who got lost in a cave and spent five weeks there without food and water before he was found by his rescuers. During this ordeal his first (conscious) reaction was to actively search for food — due to orexin, a hormone produced in the hypothalamus; orexin is normally generated to trigger alertness and all parts of the body to work faster. However, at a later stage, some 'more hardwired' instincts (inherited by humans from more primitive species through evolution) started to prevail in the brain and everything slowed down to ensure survival when energy sources became short. There is a video about this case on YouTube that can be accessed from this website: http://videos.howstuffworks.com/discovery/6835-human-body-built-for-survival-video.htm

Secondly, a good illustration of where instincts rest in humans is provided by Paul McLean's triune model. The model states that the human brain has *three* independent (and behaviourally concurrent!) brains, which were developed successively in response to evolutionary needs. They are reptilian (responsible for survival), paleomammalian or limbic (responsible for emotions) and neomammalian or neocortex (responsible for higher-order thinking). The lowest one, the reptilian brain (or R-complex), is the one which is inherited from reptiles. This is where our instincts rest. This brain is active all the time even in deep sleep. We do not sense this reptilian brain in our consciousness under normal conditions. However, in the conditions like those of Jean-Luc Josuat's ordeal the R-complex takes control of our bodies to help them survive.

So in this chapter, we strongly hypothesise that the manifestation of these different brains is driven by the energy levels in the body, and with this hypothesis we enter the cyber-world and think of electronic systems of the future — with the idea of Darwinian evolution also being transferred to the cyber-world.

13.3. Survival and Survivability in Electronic Systems

Let's now turn our attention to artificial systems, like information systems, and raise two key questions about survival: 'survival from what?' and 'survival of what?' First of all, let's see what sort of 'disasters' we should imagine that the systems would need to survive. We can roughly categorise them into the following three groups:

(1) Faults and degradation *inside* the system: defects, ageing, transients (inside gates, crosstalk on signal lines, IR drops).
(2) Upsets *outside* the system: radiation, power supply drops, signal distortions.
(3) Miscellaneous physical effects (*both internal and external*): temperature fluctuations, electro-magnetic interference.

Now, what aspects of the systems can we consider for survival? They are mainly, but not exclusively: structure, behaviour, and specific (or purposeful) functionality (defined by the system's user for example).

Combining the sources of impairments and their effects on the system, one would conventionally consider ways of how the system would react to them. Here, the reader might see some relationships if not similarities between the property of survivability and following properties, sufficiently well explored in the ICT domain: tolerance, resilience, recoverability, longevity etc. (It is very tempting to start thinking about such even 'more biological' properties such as reproducibility, especially if our notion of survival may one day stretch to thinking about genetics and preservation of species — well, in a few years with the developments in DNA computing we may have a chance!) Let's, at least, briefly contrast survivability with two fairly common properties:

- Dependability (Fault-tolerance):
  Dependable systems typically want to restore their full functionalities, hence they have large costs for redundancy; survivability is supposed to be less resource-demanding, or in other words the system may continue to work even with incomplete power levels.
- Graceful degradation:
  Gracefully degrading systems typically have a smooth (often quantitative) reduction in their performance (cf. today people talk

about approximate computations and trade-offs between accuracy and quality of service), rather than 'qualitative' transitions to a more restricted (more critical) set of functionalities as needed for survival.

From these two brief comparisons we can see that the key difference between survivability and other seemingly similar properties lies in the way we approach the energy aspect. We start to talk about survivability when the system's power is variable, intermittent, sporadic etc. Of course, the scale and range of power and energy disruptions would matter here as well, but in our simple approximation, the notion of survivability, similar to biology, refers first of all to the power conditions. For years, ICT systems have been designed to be fault-tolerant, robust and resilient to faults, ageing etc., but they have always been assumed to be fully powered. Of course, otherwise, how can one activate the fault-detection and correction procedures and engage recovery mechanisms.

At this point, however, the reader might actually stop us by saying that survivability has been studied in ICT. Indeed, it has — but conventional survivability in ICT is more about software systems (cf. [Ref. 2]) that make transitions between different services depending on the operating environment.

What we are interested in here is different. It is what we call '*Deep, or Instinct-based, Survival*', as opposed to conventional survivability, where again, as it is about software, there is very little scope to think about serious power-related issues, such as power deficiency or interruptions.

So, conventional survivability does not consider deep, embedded layers of hardware/software that work in proportion to the level of available energy/power resources. Thus, *Deep Survival* is a new concept, inspired by nature, which maintains operation in several structural and behavioural layers, with mechanisms ('instincts') developed and accumulated in bodies due to biological evolution. So, we end this section by postulating that survivability cannot be achieved in the system without providing it with sufficient back-up in the form of instinct. And, as we can see from our quotations of Darwin and Gibbon, we must really talk about an independent *layer* of activity in the system's structure, so independent that even the ways of its powering are independent of those of the 'reasoning' layers. We will therefore have to first look at how power may modulate the system's functionality, the subject of our next section.

13.4. Power-Modulated Computing and Functionality Layers

In this paper we postulate that the principle of *power (energy)-modulated computing* [Ref. 3] is fundamental for deep survival. In other words, until and unless we start designing systems in such a way that the incoming power is actually the driver of the functional behaviour we will not be able to build systems that can survive. Yet, putting it even more strongly, until we only limit our design approaches to power-efficiency rather than power-modulation, our systems will not be fully survivable. Here are some further arguments in favour of this view.

Any piece of electronics becomes active and performs to a certain level of its delivered quality in response to some level of energy and power. A quantum of energy when applied to a computational device can be converted into a corresponding amount of computation activity. Depending on their design and implementation, systems can produce meaningful activity at different power levels. As power levels become uncertain we cannot always guarantee completely certain computational activity. Good characterisation of power profiles for the system in space and time is important for designing systems for survival. Figure 13.1 illustrates this idea.

At any moment in time we have a determinate trace of power supply in the past but the future is indeterminate. The system, thanks to its sensing abilities and initial forms of intuition, can make some localised prediction from every moment at present, and its ability to compute (say in terms of the rate of activity) will be determined by the actual power levels. This brings us to the link

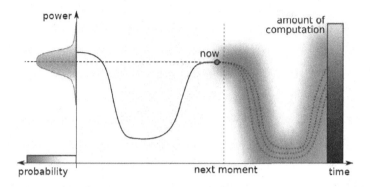

Fig. 13.1. Power profile in time, its uncertainty and illustration of power-modulated computing.

with the recently published ideas of power-proportional computing [Refs. 3 and 4]. Power proportionality, however, has two forms. One, more conventional, form concerns the fact that the system is power-proportional when its power consumption is proportional to its service demand. When systems are driven by the service demand they tend to follow the principle of multi-modality, where the system 'consciously' switches between a full functionality mode to a hibernating mode primarily depending on the data-processing requirements. Survival aspects here are limited to the ability of mode management.

But what if the power level drops? Here we are faced with the second form of power-proportionality, which in our view lends itself to a more general form of survivability. To extend the frontier of survivability, system design should also follow the *power-modulation approach*, and this leads to structuring the system design along partially or fully independent layers (cf. Darwin's "The very essence of an instinct is that it is followed independently of reason.").

Multiple layers of the system architecture can turn on/off at different power levels (cf. analogies with living organisms' nervous systems, or underwater life, or layers of expensive/cheap labour in most of the resilient economies). As power goes lower higher layers turn off, while the lower layers ('back up') remain active — this is where instincts become more in charge!

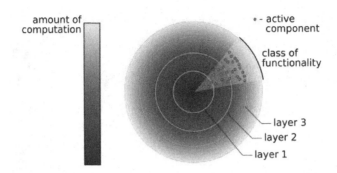

Fig. 13.2.   Layered computational activity in response to power levels.

The more active layers the system has, the more resourceful and capable of surviving it is. This layered view is reflected in Fig. 13.2, which puts it in analogy with the sea layers and ability of different forms of life to survive in different conditions of sunlight penetration.

Figure 13.3 illustrates the difference between traditional and energy-modulated system design. In the next section we will attempt to present our list of most basic instincts that the system needs to maintain for survivability.

13.5. Basic Instincts: Self-Awareness and New Sensing

The following categories of instincts can be identified in electronic computer systems that can help them to be better equipped for survival. The most important is probably energy/power-awareness, i.e. sensing, detection and prediction of power failures. The next one is the ability of storing energy 'for a rainy day'. Other instincts involve mechanisms for retaining key data, reactive and optimising mechanisms, and layers of power-driven functionality.

These instincts cannot work without the following basic abilities and associated actions:

- ability to accumulate *some* energy, initially and at any time after long interruption, say by charging a passive element;
- ability to switch, e.g. generate events;
- ability to decide, e.g. whether there is an event or not.

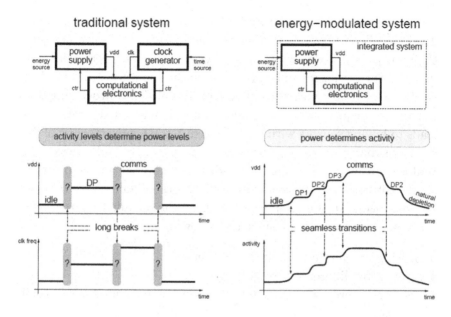

Fig. 13.3. Traditional versus energy-modulated design.

These actions underpin two major categories of instinct-supporting mechanisms:

- Mechanisms in energy and data processing domains:
  - Reference-free self-sensing and monitoring [Refs. 5 and 8],
  - Retention memory for survival [Ref. 9],
  - Elastic power-management for survival [Ref. 10].
- Mechanisms in communication fabric:
  - Monitoring progress in transactions (link level failures, deadlock detection) [Refs. 11 and 12],
  - Power noise and thermal monitoring [Ref. 13],
  - Non-blocking communications [Ref. 14].

In this chapter we restrict ourselves by discussing only the first category of mechanisms. An interested reader may find the description of mechanisms in the second category in our papers [Refs. 11–15]. Our main focus here in on self-awareness, hence sensing is our priority. Sensors must work in changing environments with uncertainty, where constant and reliable references are not available. Traditionally, sensors used in electronic systems are quite heavy — their purpose is to convert some physical form of information into digital form so that it can be processed in the computing system. Normally, this is done with the purpose of digital signal processing with fairly high requirements for fidelity and signal-to-noise ratio. This leads to having sensors with fully fledged A-to-D converters involving accurate voltage or time references supplied from outside. In systems that are autonomous and in the conditions where the aim is to survive this is not possible. Hence our target is to design an entirely different sort of sensors. In this chapter we focus on the so-called *reference-free* sensors, where we will consider the following options:

- Sensing by charge-to-digital conversion;
- Sensing by differentiators in delays;
- Sensing by crossing characteristic mode boundaries such as oscillations;
- Sensing by measuring metastability rates.

All of these sensors have some digital parts whose behaviour is modulated by the voltage that they sense, and this voltage is connected to the power terminal of the digital part. In this way these sensors are inherently power-modulated. We shall now describe some of such sensors.

13.5.1. *Sensing by charge-to-digital conversion*

This method involves sampling the input signal into a capacitor in the form of its electric charge and then discharging the capacitor in such a way that its charge is converted to digital code. Basically it is inspired by the challenge of building a sensor that is powered by the energy of the sensed signal itself. So, the principle of operation of such a sensor is that energy sampled in the capacitor as charge is proportional to the sensed voltage. It is then discharged through some load registering the quantity of energy (just like in a waterwheel!). For such a load we can use a self-timed counter as shown in Fig. 13.4. The bottom part of the figure shows voltage

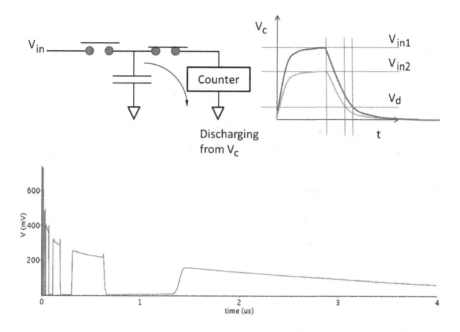

Fig. 13.4. Charge-to-digital conversion principle.

on the capacitor as a function of time. We have investigated this relationship and found that it is subject to a complex behaviour of the switching gates in the counter, which are defined by the characteristics of their constituent transistors in different modes and mechanisms, including superthreshold, subthreshold, leakage etc. Under reasonable approximations the analytical characteristic of voltage versus time is a hyperbola rather than exponential while the transistors operate in superthreshold mode [Ref. 15].

Let's now discuss the reference-free issue in this method. In the absence of external voltage and time references, we still need to control time in order to decide when to stop the discharging process while the level of voltage in the counter is sufficiently high, so the code stored in the counter can be recorded before the counter stops counting. We should stop counting irrespective of V_{in} — constant sensing/conversion delay.

However, this 'same time' implies timing reference or some clock. Hence we need to produce a voltage level V_d such that is a constant

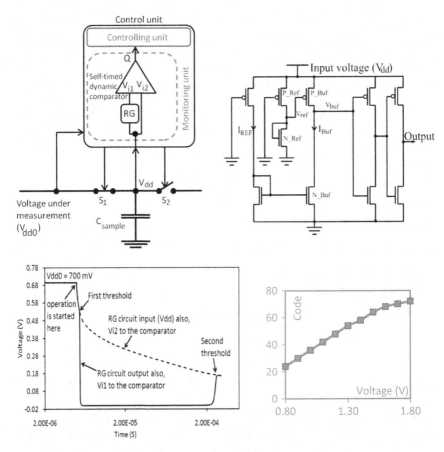

Fig. 13.5. Sensor control and its internal reference generator, the timing controlled by RG and the measured code vs input voltage (data from the fabricated 180 nm chip).

reference. V_d could be based on some internal constant such as the threshold of a transistor (similar to the idea of bandgap).

The circuit shown in Fig. 13.5 illustrates how the control circuit and internal reference generator can be built. The waveform in this figure shows that the event of crossing the second threshold corresponds to stopping the counting and latching the code from the counter. We have designed and fabricated a sensor chip in 180 nm TSMC via Europractice. We connected the chip to a 10 nF sampling capacitor and tested the sensor — the results are plotted in the above figure.

This experiment has shown the feasibility of building a sensor that is powered by the signal it senses and that is reference-free. In the following sections we will show ideas for building sensors that can be used in the highly variable conditions. We have not yet brought them to the same level of experimental implementation as the above sensor, but there are plans to do so.

13.5.2. *Sensing by delay differentiators*

The idea of sensing using delay differentiators is as follows. We need to design two circuits, which can operate in a range of voltages of our interest. The circuits, however, must have their delays scaled differently to the supplied voltage, as shown in Fig. 13.6 (left-hand side). If this is the case, then the difference between these delays will represent some characteristic form of (ideally, proportional in some critical range of interest) dependence on the supply voltage. The right-hand side of Fig. 13.6 shows that the digital value of the measured voltage can be obtained by measuring the time when Circuit 1 finishes against Circuit 2. We thus need a mechanism of registering the position of where the signal is in Circuit 2 when Circuit 1 is finished.

For example, we have observed the difference (mismatch) in delay scaling between SRAM cells and logic gates, as shown in Fig. 13.7 This mismatch rapidly increases (in terms of the number of inverters that need to match the delay of the SRAM cell, which acts as Circuit 1) when V_{dd} drops below 0.7 V (for 90 nm technology). Now,

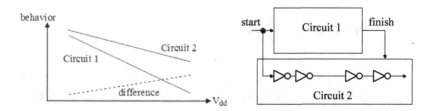

Fig. 13.6. Principle of delay difference based sensing.

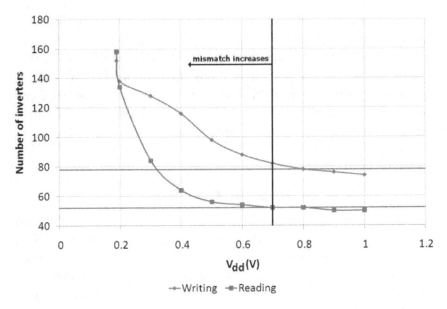

Fig. 13.7. Mismatch between inverter chain and memory cell delay (90 nm technology).

replacing the line of inverters with a self-timed counter (similar to the one used in the charge-to-code converter) to act as Circuit 2, which is started together with the SRAM cells and stopped when the reading (or writing) of the cell finishes (we used a self-timed SRAM with explicit completion detection), allows us to register the binary code for the delay difference. This is shown in Fig. 13.8 on the basis of spice simulations for a 90 nm technology node. Although the linearity of this sensor is quite limited, it can still be used for the purposes of condition monitoring we are interested in and what's really important is that it is completely reference-free.

13.5.3. *Sensing by oscillation detection*

It is often the case that we need to sense voltage, say power supply, only to the point where it crosses certain level, for example, the level at which some 'reasoning' parts of the system can no longer be trusted. This kind of sensing can be done with a circuit which changes its operating mode, for example, from stable to oscillatory.

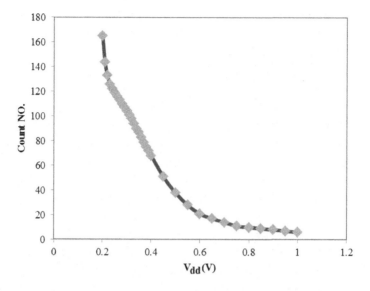

Fig. 13.8. Voltage sensing result using memory-logic mismatch.

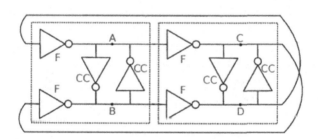

Fig. 13.9. Voltage-modulated oscillator.

An example of such a threshold-crossing oscillator is shown in Fig. 13.9. It consists of two stages, each containing a pair of forward (F) inverters and a pair of cross-coupled (CC) inverters. The circuit has two operating modes: oscillation and latching/locking. When the supply voltage V_{dd} drops below the certain V_{thr} level the circuits oscillates, as shown in Fig. 13.10.

The specific value of V_{thr} can be defined at design time by setting the ratio between transistor sizes:

$$r - \frac{width \ of \ CC}{width \ of \ F}.$$

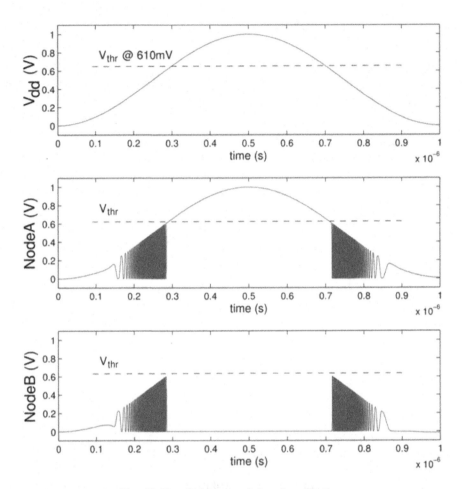

Fig. 13.10.   Voltage-modulated oscillation.

The effect of the ratio on the V_{thr} is shown in Fig. 13.11. The overall setup for detecting an event of V_{thr} crossing via oscillation is shown in Fig. 13.12. It uses a self-timed counter, initially reset to zero but introduces some delay of counting until the most significant bit is set to 1, to guarantee that the oscillations are stable. As before, it is easy to see that this method of sensing is free from external references. The behaviour is completely determined by the internal characteristics of the devices.

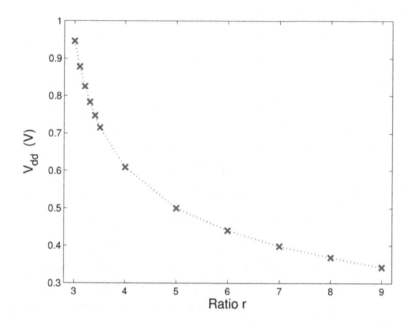

Fig. 13.11.   Transistor size ratio vs V_{dd} at which the circuit oscillates.

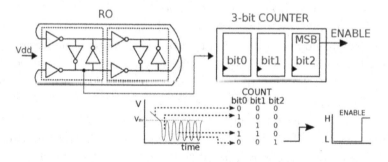

Fig. 13.12.   Setup for oscillation-based sensing.

13.5.4. *Sensing by measuring metastability rates*

Finally we present another technique for voltage sensing (it is also applicable to temperature sensing). It is based on the use of metastability in bistable devices. Metastability offers a nice way of removing external references in the voltage and temperature sensor.

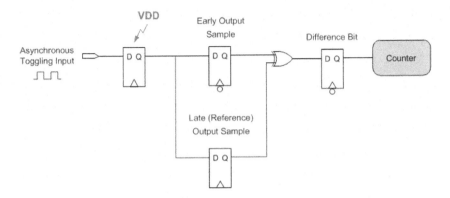

Fig. 13.13.   Circuit for metastability rate measurement.

When the setup and hold time conditions of a flip-flop are not met, the flip-flop may become metastable. A metastable flip-flop will take extra time to decide whether to go logic high or low (decision time = clock-to-q delay). The 'decision making' time constant (τ) is a function of V_{dd}. So, the idea of the method is to use the time constant (τ) to quantify V_{dd}. What we need to do is to count the rate at which the flip-flop fails to decide!

The sensor circuit shown in Fig. 13.13 works as follows. Firstly, the left-most flip-flop (call it FF1) often becomes metastable because its input is asynchronous. Secondly, when FF1's output is delayed, the early and late samples of FF1's output (captured at the following falling and rising edges respectively) will be different. Finally, the counter counts these instances. Its output after a fixed period of time is an exponential function of the time constant τ, which is determined by the sensed parameter. The advantages of this method are that it is purely digital, very compact and offers sufficiently high precision. We proved this concept in FPGA (Altera Cyclone II), and the results are shown in Fig. 13.14 (in a semi-logarithmic plot).

13.6.   Elastic Memory for Data-Retention in Instincts

We now illustrate a way of designing retention storage (SRAM) for survival. We call it elastic because it is completely self-timed and operates correctly in a wide range of supply voltages, both stable

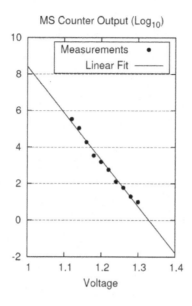

Fig. 13.14. Voltage sensing results for the FPGA prototype.

and time-varying. This SRAM can be built around different types of cells; for example, we have designs for 6T and 10T cells. One can use a 6T solution for energy-efficiency and 10T for core-function survivability. One can build control for such an SRAM array with different types of completion detection, again depending on the need to mitigate variation between columns. For example, a version with more economic completion detection (data bundling) is shown in Fig. 13.15.

A *speed-independent control* circuit for the SRAM is shown in Fig. 13.16. The timing diagram shown in Fig. 13.17 shows the simulation trace for the SRAM as it works for Data Write with the time varying V_{dd} supply. It is easy to notice the response time with which the memory sends the Wack signal is modulated by the V_{dd} (for smaller V_{dd} the delay between Wreq and Wack is longer).

For a 6T case we have built an ASIC prototype to prove the concept. The layout of the die is shown in Fig. 13.18. The chip was successfully tested and one of the traces of the Wack signals captured by oscilloscope clearly shows the effect of the switching behaviour

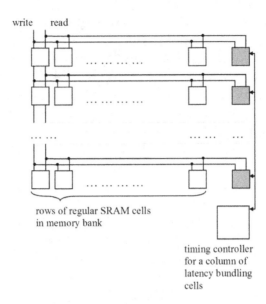

Fig. 13.15.   SRAM array with data bundling.

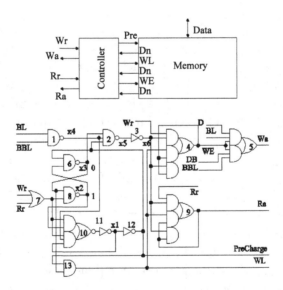

Fig. 13.16.   Speed-independent control circuit for SRAM.

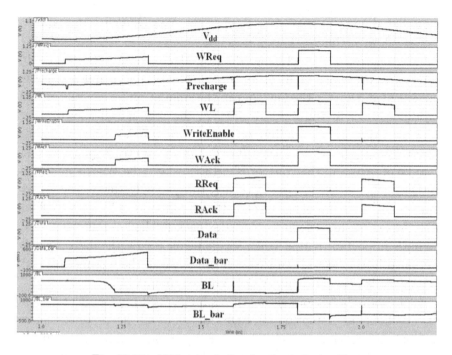

Fig. 13.17. Write simulation for time-varying V_{dd}.

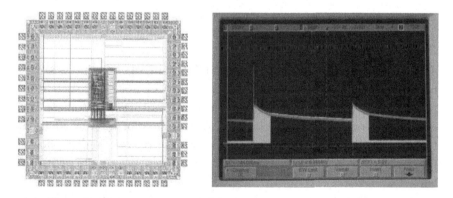

Fig. 13.18. SRAM chip layout (UMC 90 nm Europractice) and control signal trace for varying V_{dd}.

modulated by V_{dd} (one can observe the V_{dd} changing in a quick-charge-slow-discharge shape).

During testing the chip we discovered interesting effects of self-timed SRAM which confirm its time elasticity and useful properties for survival. Despite the fact that in the simulation we saw the circuit working down to the level of 190 mV, the real silicon showed that the SRAM worked steadily for V_{dd} above 0.75 V, after which its control logic 'froze' in either its setting or resetting phases. This can be observed in the trace of Fig. 13.18 where the Wack signal gets 'stuck' either in the low or high state. Interestingly, due to the speed-independent nature of the circuit, the circuit smoothly recovers from the 'frozen' state as soon as V_{dd} goes back to the level above 0.75 V. What's important is that when V_{dd} is below 0.75 V the data is safely retained in the SRAM (this was checked during the testing process). The data is retained while V_{dd} is greater than 0.4 V.

The above behaviour shows that a fully speed-independent SRAM is excellent as retention storage for survival in power-deficient regimes. It provides self-detection of the power condition by 'freezing', an early warning, well before the system starts to lose its data.

13.7. Retaining Energy: Elastic Power Management for Instincts

The design of ICT systems destined for survival will increasingly be more holistic and will have to take care of not only their data processing parts, such as sensing and computing electronics, but also their power supply electronics. We are exploring new ideas in this direction. They involve more active use of switched capacitor circuits for DC/DC conversion. Conventionally there are switched capacitor DC/DC converters (SCCs). They convert constant input V_{dd} to constant output V_{dd} according to a set of ratios. However, SCCs usually rely on the availability of stable sources of time and voltage references. Instead, under harsh operating conditions such references may not be available. Hence, we develop a different type of switched capacitor circuits that are aware of the presence of self-timed circuits as their load. We call them capacitor bank blocks (CBBs). We have also designed hybrid CBBs that can work as SCCs and CBBs depending on the conditions and whether the load electronics is synchronous or asynchronous. Details of this method can be found in [Ref. 10].

13.8. Conclusions and Outlook

As stated in the abstract and introduction, this chapter was inspired by the ideas of incorporating self-awareness into systems that have been studied by Professor Cheung in the context of improving the performance of electronic systems under process variations and ageing. We take self-awareness further, and with the help of biological analogy, consider survival instincts here. The paper has focussed almost exclusively on the techniques and examples of circuits for survivability that support an 'instinct layer', which is supposed to remain alive and operational under the conditions of power instabilities and lack of power.

We are currently involved in an EPSRC-funded project 'Staying alive in variable, intermittent, low-power environments' (SAVVIE), in collaboration with Dr. Bernard Stark of University of Bristol. The project's main aim is to develop techniques for enabling systems to survive in the top left corner of the energy-power state space depicted in Fig. 13.19. While there exist methods that support trajectories like T1 and T2 in this state space, approaches to cater for trajectories such as T3 and T4 are in their infancy. We hope that the ideas that have been described in this chapter will contribute to this aim.

A list of outgoing research directions we are currently pursuing:

- More diversification — power and data processing paths intertwined, mixed digital and analogue fabrics, synchronous and asynchronous fabrics, multiple technology fabrics.
- New modelling and design approaches — models that capture multi-modal and multi-layer architectures; combining structure and behaviour in models, capturing overlay in functionality.

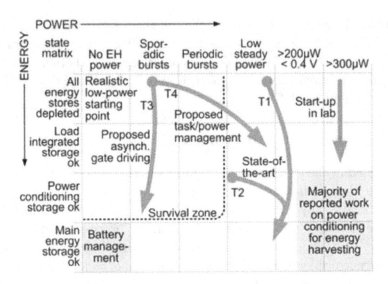

Fig. 13.19.   Energy-power state space in the SAVVIE project (courtesy of B. Stark).

There is plenty to investigate on this path, and the research is already under way at Newcastle and in collaboration with our partners from Southampton, Imperial and Manchester under the programme grant PRiME that will explore energy-reliability tradeoffs in designing future many-core embedded systems.

Acknowledgments

This chapter would not have been written without the combined effort of the author's research team, the Microelectronics Systems Design group at Newcastle. The list of people actively working in this area can be found on the group's webpage: http://async.org.uk.

Our collaboration with Dr Bernard Stark's team at Bristol in the SAVVIE project and our recent collaboration with Universities Southampton, Bristol and Imperial College London under the Holistic project (http://www.holistic.ecs.soton.ac.uk/) are acknowledged with deep gratitude.

References

1. J. M. Levine *et al.* Online Measurement of Timing in Circuits: for Health Monitoring and Dynamic Voltage & Frequency Scaling, *in Proc. International Symposium on Field-Programmable Custom Computing Machines*, pp. 109–116, 2012.

2. J.C. Knight and E.A. Strunk. Achieving Critical System Survivability through Software Architectures, in *Proc. Architecting Dependable Systems II*, pp. 51–78, 2004.

3. A. Yakovlev. Energy-Modulated Computing, in *Proc. Design, Automation & Test in Europe Conference & Exhibition*, pp. 1340–1345, 2011.

4. R. Ramezani *et al.* Energy-modulated Quality of Service: New Scheduling Approach, in *Proc. Faible Tension Faible Consommation*, pp. 1–4, 2012.

5. R. Ramezani *et al.* Voltage Sensing Using an Asynchronous Charge-to-Digital Converter for Energy-Autonomous Environments, *IEEE Journal on Emerging and Selected Topics in Circuits and Systems*, 3(1), 35–44, 2013.

6. D. Shang, F. Xia and A. Yakovlev. Wide-Range, Reference Free, On-chip Voltage Sensor for Variable Vdd Operations, in *Proc. International Symposium on Circuits and Systems*, pp. 37–40, 2013.

7. G. Tarawneh, T. Mak and A. Yakovlev. Intra-chip Physical Parameter Sensor for FPGAs using Flip-Flop Metastability, in *Proc. International Conference on Field Programmable Logic and Applications*, pp. 112–119, 2012.

8. I. Syranidis, F. Xia and A. Yakovlev. A Reference-free Voltage Sensing Method Based on Transient Mode Switching, in *Proc. Conference on Ph.D. Research in Microelectronics and Electronics*, pp. 1–4, 2012.

9. A. Baz *et al.* Self-timed SRAM for Energy Harvesting Systems, *Journal of Low Power Electronics*, 7(2), 274–284, 2011.

10. X. Zhang *et al.* A Hybrid Power Delivery Method for Asynchronous Loads in Energy Harvesting Systems, in *Proc. International NEWCAS Conference*, pp. 413–416, 2012.

11. L. Dai *et al.* Monitoring Circuit Based on Threshold for Fault-tolerant NoC, *Electronics Letters*, 46(14), 984–985, 2010.

12. R. Al-Dujaily *et al.* Embedded Transitive Closure Network for Runtime Deadlock Detection in Networks-on-Chip, *IEEE Transactions on Parallel and Distributed Systems*, 23(7), 1205–1215, 2012.

13. N. Dahir *et al.* Minimizing Power Supply Noise through Harmonic Mapping in Networks-on-Chip, in *Proc. International Conference on Hardware/software Codesign and System Synthesis*, pp. 113–122, 2012.

14. F. Xia *et al.* Data Communication in Systems with Heterogeneous Timing, *IEEE Micro*, 22(6), 58–69, 2002.

15. R. Ramezani and A. Yakovlev. Capacitor Discharging through Asynchronous Circuit Switching, in *Proc. International Symposium on Asynchronous Circuits and Systems*, pp. 16–22, 2013.

Index

Printed in the United States
By Bookmasters